Practical Handbook of Remote Sensing

The number of Earth observation satellites launched in recent years is growing exponentially, along with the datasets they gather from free-to-access and commercial providers. The second edition of *Practical Handbook of Remote Sensing* is updated with new explanations and practical examples using the Copernicus satellite data and new versions of the open-source software. A new chapter and new applications have also been added. Thoroughly revised, the handbook continues to be a practical "how-to" remote sensing guide for those who want to use the technology, understand what is available, how to access it, and answer questions about our planet but do not necessarily want to become scientific experts.

- Updated with recent changes and developments in the Earth Observation industry.
- Updated to reflect the latest software and data versions, making them easy to use.
- Introduces the Copernicus missions and gives users practical examples of how to find, download, process, and visualize this free-to-access data.
- Includes a new chapter on atmospheric remote sensing extending the examples to atmospheric and climate applications.
- Brings in the latest and foreseen future scientific and technical developments.

This book is intended for those with no prior knowledge of remote sensing but who need to use the technology in their work, including professionals in administration, urban planning, environmental management, and natural disaster. Undergraduate geography and geosciences students who need to study introductory remote sensing will also find this book a solid foundation for their studies and citizens interested in understanding the environment and what can be detected from space.

Practical Handbook of Remote Sensing

Second Edition

Samantha Lavender
Andrew Lavender

CRC Press
Taylor & Francis Group
Boca Raton London New York

CRC Press is an imprint of the
Taylor & Francis Group, an **informa** business

Designed cover image: Credit for the compilation of the front cover image goes to Samantha Lavender; individual images are (top left: Data courtesy of NASA/USGS, bottom left: Data courtesy of U.S. Geological Survey, top right: Data courtesy of NASA/USGS/Copernicus/ESA, middle right: Data courtesy of Copernicus/ESA, bottom right: Data courtesy of USGS/NASA.) Individual images cited within the work.

Second edition published 2023
by CRC Press
6000 Broken Sound Parkway NW, Suite 300, Boca Raton, FL 33487-2742

and by CRC Press
4 Park Square, Milton Park, Abingdon, Oxon, OX14 4RN

CRC Press is an imprint of Taylor & Francis Group, LLC

© 2023 Taylor & Francis Group, LLC

First edition published by CRC Press 2017

Library of Congress Cataloging-in-Publication Data
Names: Lavender, Samantha, author. | Lavender, Andrew, author.
Title: Practical handbook of remote sensing / Samantha Lavender, Andrew Lavender.
Description: Second edition. | Boca Raon, FL : CRC Press, 2023. |
Includes bibliographical references and index. |
Summary: "The number of Earth observation satellites launched in recent years is growing exponentially, along with the datasets they gather from free-to-access and commercial providers. The second edition of Practical Handbook of Remote Sensing is updated with a new chapter, new explanations and practical examples using the Copernicus satellite data and new versions of the open-source software. Thoroughly revised, and intended for those with no prior knowledge of remote sensing, the handbook continues to be a practical "how-to" guide. Professionals in administration, urban planning, environmental management, and students of geography and geosciences, will find this book helpful"– Provided by publisher.
Identifiers: LCCN 2022049011 (print) | LCCN 2022049012 (ebook) |
ISBN 9781032223582 (hardback) | ISBN 9781032214337 (paperback) | ISBN 9781003272274 (ebook)
Subjects: LCSH: Remote sensing–Handbooks, manuals, etc.
Classification: LCC G70.4 .L38 2023 (print) | LCC G70.4 (ebook) |
DDC 621.36/78–dc23/eng/20221215
LC record available at https://lccn.loc.gov/2022049011
LC ebook record available at https://lccn.loc.gov/2022049012

ISBN: 978-1-032-22358-2 (hbk)
ISBN: 978-1-032-21433-7 (pbk)
ISBN: 978-1-003-27227-4 (ebk)

DOI: 10.1201/9781003272274

Typeset in Palatino
by codeMantra

Contents

List of Figures

List of Tables

Preface

People see remote sensing data every day without realizing it, as it's included in news programs and weather forecasts, and some regularly use it with software packages such as Google Earth. Many people don't realize they're looking at satellite data, and even fewer realize that these data are often freely available and accessible to anyone.

Every day, hundreds of satellites orbit the Earth, and many are collecting environmental data used to understand both the short- and long-term changes to the planet. However, you need a little bit of knowledge to find, download, analyze, and view the data. This is where this book comes in.

I've been involved in remote sensing for more than 30 years, and I've written several chapters for academic textbooks. These books are often only understandable by the people who already know the subject. From the very start, the aim of this book was different. I wanted to create a book that could take a complete beginner through the basic scientific principles and teach them to do their own practical remote sensing at home or work, using just a personal computer.

I realized quite quickly that if I was going to write a "how-to" book for people without any experience in remote sensing, I needed someone who knew nothing about the subject to act as the tester; this was where my husband came in!

So as the expert, and nonexpert, we've written this book together. It's been an interesting, challenging, occasionally argumentative but ultimately pleasurable experience, as we searched for the compromise between scientific theory and understandable language. As a result, we've used simplified explanations and only a few equations, rather than trying to explain the full complexity. We hope this balance will appeal to those new to the subject and those used to reading technical documents.

In this second edition, we have updated the descriptions and practicals to include more recent data sources, including the Copernicus Sentinel satellites and the increasing array of commercial CubeSat satellites. The first three chapters give a preliminary introduction to remote sensing, how it works, and the available data. Then, Chapters 4–7 cover the theory and application of basic technical remote sensing skills, using free downloadable software, guiding you through finding, downloading, manipulating, and viewing data from the Landsat and Sentinel satellites. We've chosen these two sets of satellites as the main demonstrators, as they have global free-to-access archives of data and it's relatively easy to handle in the open-source software we suggest.

Chapters 8–12 focus on a series of remote sensing applications, where the data are used to research, monitor, and solve real-life challenges. We start with the urban environments and move on to the evolution of the natural landscape, followed by the terrestrial water cycle, coastal waters, before finishing off with atmospheric gases and pollutants. Chapter 13 concludes the book by considering the future of remote sensing, alongside how readers can go on to develop their own interests and skills.

We hope this approach makes the contents accessible, and we'll be interested to hear feedback. As the software and data sets used in the book are continually changing, and to provide an interactive environment, it's accompanied by a learning resource website (https://www.playingwithrsdata.com/).

Overall, we feel that the book has remained true to the original thoughts and hope those working their way through will gain a glimpse into the complexity alongside the potential for remote sensing. We think that those who read this book will undertake and continue remote sensing activities for themselves and put this valuable societal resource to greater use, and hopefully to a broader audience.

Samantha Lavender

Acknowledgments

We acknowledge those in the academic and commercial remote sensing communities who continue to work on developing both satellite missions and applications, as this book showcases just a tiny fraction of what's possible.

The figures have come from freely available data sets, funded through space agencies such as European Space Agency (ESA) and National Aeronautics and Space Administration (NASA), alongside the Copernicus Programme and organizations such as the National Oceanic and Atmospheric Administration (NOAA) and the United States Geological Survey (USGS). We've also used example commercial data sets from the Japan Aerospace Exploration Agency (JAXA) and Airbus Defence and Space, which provide insight into this additional resource.

The practical exercises primarily use two open-source packages. Sentinel Application Platform (SNAP) is a set of toolboxes for scientifically exploiting the Sentinel missions developed by ESA. Quantum Geographic Information System (QGIS) is licensed under the GNU General Public License, which is an official project of the Open-Source Geospatial Foundation. We've also included the QGIS Semi-Automatic Classification Plugin, developed by Luca Congedo, and took inspiration from the accompanying tutorials available through a Creative Commons License.

Authors

Samantha Lavender, PhD, has more than 30 years of remote sensing research experience, focusing on using Earth observation to help answer questions about our planet's resources and behavior. After earning a PhD, she focused on the remote sensing of the Humber plume using airborne data and was a researcher at the Plymouth Marine Laboratory and then a lecturer at the University of Plymouth. In 2012, with Andrew, she formed Pixalytics Ltd and is the Managing Director of this commercial remote sensing company. Dr. Lavender is also an Associate Editor for Remote Sensing of Environment and has an ongoing active interest in research alongside widening the community using remotely sensed data. She has previously served the community as chairman of the British Association of Remote Sensing Companies (BARSC), chairman of the Remote Sensing and Photogrammetry Society (RSPSoc), and Treasurer for the European Association of Remote Sensing Laboratories (EARSeL).

Andrew Lavender is a novice remote sensor. Having founded Pixalytics Ltd alongside his wife, he primarily focuses on the company's administrative and social media activities. Andrew occasionally undertakes modest aspects of simple remote sensing to produce marketing images and acts as the 'future reader' of the book, including testing all of the exercises. Outside of Pixalytics, he is a short story and flash fiction writer and teaches creative writing.

List of Symbols

Symbol, followed by description, unit, and the section where it was first used in the book:

b	Wien's displacement constant $2.8977685 \pm 51 \times 10^{-3}$ (m · K) – 8.4
$B_\lambda(T)$	brightness temperature (K) at a particular wavelength – 8.5
C_1	first radiation constant (Wm^2) – 8.5
C_2	second radiation constant (m · K) – 8.5
E_d	solar irradiance (Wm^{-2}) – 8.3
F_b	emitting radiant flux of the blackbody (Wm^{-2}) – 8.4
F_r	emitting radiant flux of the real material (Wm^{-2}) – 8.4
K_1	first thermal band calibration constant (Wm^{-2}/sr/µm) – 8.5
K_2	second thermal band calibration constant (K) – 8.5
K_d	diffuse attenuation coefficient (m^{-1}) – 11.4
L	radiance (Wm^{-2}/sr/µm) – 8.3
L_λ	radiance at a particular wavelength (Wm^{-2}/sr/µm) – 8.5
L_g	ground radiance (Wm^{-2}/sr/µm) – 8.3
L_s	sensor radiance – 8.3
$L_w(\lambda)$	water-leaving radiance (Wm^{-2}/sr/µm) – 11.1
R	reflectance (unitless, expressed as a number between 0 and 1) – 8.4
$R_{rs}(\lambda)$	remote sensing reflectance (sr^{-1}) – 11.1
T	temperature (K) – 8.4
ε	emissivity (unitless, expressed as a number between 0 and 1) – 8.4
λ	wavelength (nm) – 8.5
λ_{max}	wavelength of peak radiation (m, which can be converted to µm) – 8.4

List of Acronyms and Abbreviations

2D	two-dimensional
3D	three-dimensional
3U	three CubeSat units
AC	Atmospheric Correction
AI	Artificial Intelligence
Airbus DS	Airbus Defence and Space
AMSR-E	Advanced Microwave Scanning Radiometer-Earth Observing System
AOT	Aerosol Optical Thickness
ARD	Analysis Ready Data
ASI	Agenzia Spaziale Italiana
ASTER	Advanced Spaceborne Thermal Emission and Reflection Radiometer
ATSR	Along Track Scanning Radiometer
BOA	Bottom of Atmosphere
C3S	Copernicus Climate Change Service
C ID	Classification ID
C Name	Classification Name
CALIPSO	Cloud-Aerosol Lidar and Infrared Pathfinder Satellite Observation
CAMS	Copernicus Atmosphere Monitoring Service
CAVIS	Clouds, Aerosols, Water Vapor, Ice, and Snow
CCI	Climate Change Initiative
CDOM	Colored Organic Dissolved Matter
CEDA	Centre for Environmental Data Archival
CEOS	Committee on Earth Observation Satellites
CFC	Chlorofluorocarbon
CHEOS	China High-resolution Earth Observation System
CHIME	Copernicus Hyperspectral Imaging Mission for the Environment
Chlor-a	Chlorophyll-a
CIMR	Copernicus Imaging Microwave Radiometer
CLASS	Comprehensive Large Array-Data Stewardship System
CNES	Centre National d'Etudes Spatiales
CNSA	China National Space Administration
CO_2	carbon dioxide
CO2M	Copernicus Anthropogenic Carbon Dioxide Monitoring
CRISTAL	Copernicus Polar Ice and Snow Topography Altimeter
CRS	Coordinate Reference System

CSEOL	Citizen Science Earth Observation Lab
CZCS	Coastal Zone Color Scanner
DEM	Digital Elevation Model
DIAS	Data and Information Services
DLR	Deutsche Forshungsanstalt fur Luft und Raumfahrt
DMC	Disaster Monitoring Constellation
DMSP	Defense Meteorological Satellite Program
DN	Digital Number
DTM	Digital Terrain Model
EAC4	ECMWF Atmospheric Composition Reanalysis 4
ECMWF	European Centre for Medium-Range Weather Forecasts
ECVs	Essential Climate Variables
EM	Electromagnetic
EnMAP	German Environmental Mapping and Analysis Program
EO	Earth Observation
ERTS-1	Earth Resources Technology Satellite
ESA	European Space Agency
ETM+	Enhanced Thematic Mapper Plus
EUMETSAT	European Organization for the Exploitation of Meteorological Satellites
FAIR	Findable, Accessible, Interoperable, and Reuseable
fAPAR	fraction of absorbed photosynthetically active radiation
FCDR	Fundamental Climate Data Record
fCover	vegetation cover fraction
FLH	Normalized Fluorescence Line Height
FRP	Fire Radiative Power
GB	gigabyte
GCOS	Global Climate Observing System Implementation Plan
GCPs	Ground Control Points
GEOSS	Global Earth Observation System of Systems
GeoTIFF	Geostationary Earth Orbit Tagged Image File Format
GHG	Greenhouse Gas
GHRSST	Group for High-Resolution Sea Surface Temperature
GIMP	GNU Image Manipulation Program
Giovanni	Geospatial Interactive Online Visualization and Analysis Infrastructure
GIS	Geographic Information System
GLCF	Global Land Cover Facility
GloVis	Global Visualization Viewer
GMES	Global Monitoring for Environmental Security
GNSS	Global Navigation Satellite System
GOES	Geostationary Satellite system
GPM	Global Precipitation Measurement
G-POD	Grid Processing on Demand

GRACE	Gravity Recovery and Climate Experiment
GRASS	Geographic Resources Analysis Support System
H₂O	water vapor
HAPS	High-Altitude Pseudo-Satellite/High-altitude Platform Station
HF	High Frequency
HRVIR	High-Resolution Visible and Infrared sensor
ICESat-GLAS	Ice, Cloud, and Land Elevation Satellite-Geoscience Laser Altimeter System
ID	Identifier
IGS	International Ground Station
InSAR	Interferometric SAR
INSPIRE	Infrastructure for Spatial Information in the European Community
iPAR	Instantaneous Photosynthetically Available Radiation
IPCC	Intergovernmental Panel on Climate Change
IR	Infrared
ISRO	Indian Space Research Organisation
ISS	International Space Station
JAXA	Japan Aerospace Exploration Agency
L0	Level 0
L1	Level 1
L2	Level 2
L3	Level 3
L4	Level 4
LAADS	L1 and Atmosphere Archive and Distribution System
LAI	Leaf Area Index
LC	Land Cover
LEO	Low Earth Orbit
LP DAAC	Land Processes Distributed Active Archive Center
LRM	Low-Resolution Mode
LST	Land Surface Temperature
LSTM	Copernicus Land Surface Temperature Monitoring
LSWT	Lake Surface Water Temperature
LU	Land Use
LULC	Land Use and Land Cover
MB	Megabyte
MC ID	MacroClass ID
MC Name	MacroClass Name
MCI	Maximum Chlorophyll Index
MERIS	MEdium Resolution Imaging Spectrometer
MIR	Mid-Infrared
MIRAS	Microwave Imaging Radiometer with Aperture Synthesis
ML	Machine Learning

MOD04	NASA's standard aerosol product
MODIS	Moderate Resolution Imaging Spectroradiometer
MOOC	Massive Open Online Course
MSI	Multispectral Instrument
MSS	Multispectral Scanner
MTCI	MERIS Terrestrial Chlorophyll Index
NASA	National Aeronautics and Space Administration
NDVI	Normalized Difference Vegetation Index
NDWI	Normalized Difference Water Index
NIR	Near-Infrared
NISAR	NASA–Indian Space Research Organization Synthetic Aperture Radar
NO_2	nitrogen dioxide
NOAA	National Oceanic and Atmospheric Administration
NRT	Near-Real Time
NSIDC	National Snow and Ice Data Center
O_2	oxygen
O_3	ozone
OC2	Ocean Color 2
OCO	NASA's Orbiting Carbon Observatory
OGC	Open Geospatial Forum
OLCI	Ocean and Land Color Instrument
OLCI-2	Ocean and Land Color Instrument-2
OLI	Operational Land Imager
OLS	Operational Line Scan
OMI	NASA Ozone Monitoring Instrument
OSI SAF	Ocean and Sea Ice Satellite Application Facility
PALSAR-2	Phased Array Type L-band Synthetic Aperture Radar
PCA	Principal Component Analysis
PIC	Particulate Inorganic Carbon
POC	Particulate Organic Carbon
PM	Particulate Matter
PRISMA	PRecursore Iperspettrale della Missione Applicativa
QGIS	Quantum GIS
QuikSCAT	Quick Scatterometer
RA-2	Envisat Radar Altimeter-2
RBV	Return-Beam Vidicon
REDD+	Reducing Emissions from Deforestation and forest Degradation in developing countries
RMSE	Root-Mean-Squared Error
RO	Radio Occultation
ROI	Region of Interest
RSGISLib	Remote Sensing and GIS Software Library
RT	Real Time

SAFE	Standard Archive Format for Europe
SAGA	System for Automated Geoscientific Analyses
SAR	Synthetic Aperture Radar
SARIn	SAR Interferometric
SCIAMACHY	Scanning Imaging Absorption Spectrometer for Atmospheric CHartographY
SCP	Semi-Automatic Classification Plug-in
SeaWiFS	Sea-viewing Wide Field-of-view Sensor SeaWiFS
Sen4CAP	Sentinels for Common Agriculture Policy
SIRAL	SAR Interferometric Radar Altimeter
SLC	Scan Line Corrector
SLSTR	Sea and Land Surface Temperature Radiometer (SLSTR)
SMAP	Soil Moisture Active Passive
SMI	Standard Mapped Image
SMOS	Soil Moisture and Ocean Salinity
SNAP	Sentinel Application Platform
SNR	Signal-to-Noise Ratio
SPM	Suspended Particulate Matter
SPOT	Satellites Pour l'Observation de la Terre or Earth-Observing Satellites
SRAL	Synthetic Aperture Radar Altimeter
SRTM	Shuttle Radar Topography Mission
SST	Sea Surface Temperature
STAC	SpatioTemporal Asset Catalog
STEP	Science Toolbox Exploitation Platform
SWIR	Shortwave Infrared
Suomi NPP	Suomi National Polar-orbiting Partnership
Tandem-X	TerraSAR-X add-on for Digital Elevation Measurement 2011
TECIS	Terrestrial Ecosystem Carbon Inventory Satellite
TIR	Thermal IR
TIROS	Television and Infrared Observation Satellite
TIRS	Thermal Infrared Sensor
TIRS-2	Thermal Infrared Sensor 2
TM	Thematic Mapper
TOA	Top of Atmosphere
TRMM	Tropical Rainfall Measurement Mission
TROPOMI	Tropospheric Monitoring Instrument
UAV	Uncrewed Aerial Vehicles
UN	United Nations
UNEP	United Nations Environment Programme
UNESCO	United Nations Educational, Scientific and Cultural Organization
US	United States

USGS	United States Geological Survey
UTC	Coordinated Universal Time
UTM	Universal Transverse Mercator
UV	Ultraviolet
VHF	Very High Frequency
VIIRS	Visible Infrared Imaging Radiometer Suite
WGS84	World Geodetic System of 1984
WHO	World Health Organization
WMS	Web Mapping Service
WRS	Worldwide Reference System

1

What Is Remote Sensing?

Welcome to the start of your journey to remote sensing mastery! Over the next 13 chapters, we're going to take you up into space and introduce you to the world of satellite data, showing you what your eyes can see and unveiling things your eyes can't.

There are hundreds of satellites orbiting the Earth, and you use their data every day, whether it's to make mobile telephone calls, for car navigation, or to watch the news and weather. We're passionate about the satellites that collect environmental data, which can help explore, explain, and monitor what is happening here on the Earth. The data from these satellites are used to monitor climate change, the health of the oceans, provide lifesaving support in earthquakes and early warning systems for floods. All these applications, and many more, are provided through remote sensing data.

What's more, most of these data are freely available and can be used by anyone with a reasonable computer and Internet connection. This book is going to take you through the theory, with supported practical exercises, to show you how to find, download, manipulate, and view these amazing data sets. You can investigate your local area, or anywhere in the world, and improve your understanding of the environment, or perhaps even create new applications no one has developed yet. Remote sensing is a young, growing space-based industry waiting to be discovered.

This first chapter will provide you with an overview of remote sensing, by explaining what it is, its history, how it works, and why it's useful. The final section describes the structure of the book and what will be covered within the individual chapters.

We hope that the book will interest, intrigue, and inspire you to get involved with remote sensing data and begin to use it to explore our planet. Get ready to take your first step.

1.1 Definition of Remote Sensing

The simplest definition of remote sensing is being able to know what an object is without being in physical contact with it (inspired by Sabins, 1978). You do that every day with your eyes, as you don't have to touch

DOI: 10.1201/9781003272274-1

a table or a chair to know what it is. Now imagine your eyes were up in space and could see the whole world; could you tell what type of tree was in a forest by looking at it, or how warm the ocean was, or whether the level in the river is rising, or whether air quality over a particular town is good or bad? Well, this is exactly what satellite remote sensing can do.

Remote sensing is essentially the collection of data by sensors commonly on either aircraft or satellites, although other approaches are available, for example, in the Amazon, there are a number of tall vertical platforms topped with sensors that are then processed by computer systems to provide information and images about a particular area.

Because remote sensing generally monitors the planet, the term *Earth observation* (EO) has recently become popular, to describe what remote sensing does. However, there is also remote sensing of other planets, and their moons, in the solar system, and even comets, which, in 2014, was a key part of the European Space Agency Rosetta mission to land on Comet 69P Churyumov-Gerasimenko.

Remote sensing essentially gives us an opportunity to better understand what is happening on our planet.

1.2 History of Remote Sensing

Remote sensing started with cameras more than 150 years ago. In the 1840s, pictures or aerial photographs were taken from cameras secured to balloons for topographic mapping, capturing both natural and man-made features and showing the variations in the terrain. Gaspard-Felix Tournachon took photographs of Paris from his balloon in 1858, and then (in 1896) Alfred Nobel designed a system to take aerial photographs from rockets. Cameras were mounted on airplanes and became an important source of information for World War I reconnaissance and surveillance activities. This extended into space, although not yet into orbit, with V-2 rockets acquiring imagery in the mid-1940s.

The Soviet Union Sputnik 1 was the first artificial Earth satellite, which provided information on the upper atmosphere from the orbital drag and propagation of radio signals. United States (US) Military satellites (Corona program) started in 1959 with the first photographs successfully acquired from space in 1960 using cameras with film canisters, which were dropped back to Earth in reentry capsules and then caught in mid-air by airplanes. As astronauts began going into space, they also started taking photographs.

Specific developments in EO began in 1959, with the launch of the Explorer VII satellite designed to measure the amount of heat emitted and reflected by the Earth. The US meteorological satellite TIROS 1 (Television and Infrared Observation Satellite) was launched in 1960 and sent back the first satellite image of cloud patterns over the Earth.

Since then, the number of satellite missions, the data collected, and the parameters that can be derived from the data have increased significantly. Apollo 9, launched in 1968, captured the first multispectral imagery using its four-lens camera to provide photographs that were later digitized. The crew were provided with lists and maps of the target areas selected on a daily basis by the support room using weather and light predictions alongside Apollo crew reports (Lowman, 1969); before taking each photograph, the cameras were unstowed and assembled and then triggered manually at the recommended interval.

Landsat 1, originally named the Earth Resources Technology Satellite, was launched by the National Aeronautics and Space Administration (NASA) in 1972 and began the first continuous archive of EO data to support research; that's still growing today. Later on, the Nimbus-7 satellite (launched in 1978) carried the Total Ozone Mapping Spectrometer, which went on to help confirm the existence of the Antarctic ozone hole, and the Coastal Zone Color Scanner launched in 1997 was the first sensor focused on mapping the color of the marine environment. The European Union established the Copernicus EO program (https://www.copernicus.eu/en) early in the 21st century, and in conjunction with the European Space Agency, launched the first of its Sentinel satellites, the radar satellite Sentinel-1A, in 2014. To date, the program has launched eight satellites, and although primarily designed for public policymakers and public bodies responsible for environmental and security matters, this continuous archive of data is free to access to anyone globally.

The latest developments in technology surround the use of CubeSats, which are mini- or nanosatellites built from 0.1 m square cubes, with a 3U (3 × 0.1 m square unit) CubeSat having a mass of no more than 4 kg. Their size means that they're vastly cheaper to build and launch into space, and although they don't have the lifespan of larger satellites, they offer a fantastic potential to test new sensors or undertake specific short-lived measurements. These CubeSats have also enabled the creation of satellite constellations, where a large number of identical satellites orbit the Earth enabling a great frequency of data. In EO, the most famous of these constellations is from Planet Labs, which has a couple of hundred satellites and in 2017 claimed that they could capture an image of every part of the Earth's landmass every day. Outside of EO, constellations are much bigger with SpaceX's space-based Internet broadband constellation already being several thousand strong and expected to grow to tens of thousands of satellites.

Besides satellites, there are also other types of platforms (e.g., fixed-wing aircraft, helicopters, and uncrewed aerial vehicles, also popularly known as drones) that are becoming increasingly used for low-altitude imaging. Although we'll not focus on these non–space-based missions, some airborne sources of data will be mentioned as they often provide the test platforms for space-based instruments.

1.3 Principles of Remote Sensing

Remote sensing works primarily by detecting the energy reflected or emitted from the Earth as electromagnetic (EM) radiation. The EM spectrum ranges from radio waves at the longest wavelengths/shortest frequencies to microwave radiation followed by infrared, visible, and ultraviolet radiation (see Figure 1.1). These wavelengths are absorbed, and scattered, differently both within the atmosphere and when interacting with the surface of an object or region of interest (ROI). Detecting, and interpreting, the EM energy of these different wavelengths is the essence of remote sensing. There are two different approaches to doing this – passive and active remote sensing.

In passive remote sensing, the sensor detects radiation naturally emitted or reflected by the object or ROI on the Earth. Thermal remote sensing, which is in the infrared, is detected where there is a heat emission from the Earth or its atmosphere. While optical remote sensing detects the sunlight reflected off the water or land; the ratio of the light traveling downward to that being scattered back toward the sensor is termed reflectance. Another type of passive remote sensing is by microwave radiometers,

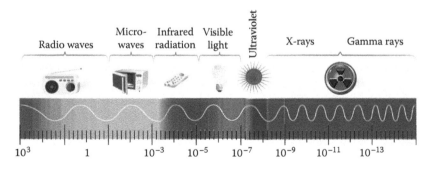

FIGURE 1.1
Electromagnetic spectrum. (Courtesy of designua © 123RF.com.)

which can be influenced by both temperature and salinity over the oceans and soil moisture over the land.

In contrast, for active remote sensing, the sensor (or platform) emits energy toward the object or ROI on the Earth and then the sensor detects and measures the strength and time delay of the return signal. An example of this method is radar (originally an acronym for RAdio Detection And Ranging) remote sensing in the microwave region. As active sensors or platforms are emitting a signal, they require more power than passive sensors and so often these satellites are only switched on when they're tasked to record data; that is, they don't routinely collect full global data sets.

Therefore, from knowing how an object or the Earth's surface interacts with EM radiation, we can infer what it is. However, the effects of the atmosphere also need to be accounted for if this is to be done accurately.

1.4 Usefulness of Remote Sensing

A key benefit of using remotely sensed data is that you don't have to go to the area being studied to sense and, hence, map it, which is particularly useful if the area is remote or difficult to get to for any reason. However, it may not mean that someone never has to go there. All remote sensing signals require interpretation. Taking measurements locally (often called ground truthing or in situ validation) means that they're compared with the satellite observations and provide a better understanding and, in some cases, improvement of the remotely sensed data products, that is, the final results after several processing steps. This is especially useful when new types of sensors have been launched or new techniques are being developed, but it is also important for assessing the stability and accuracy of an instrument, which is termed *calibration*.

The second key benefit of remote sensing is the historical archives of data that exist. The Landsat missions, which began in the 1970s, have collected a huge archive of data that are freely available for anyone to use and so can enable a historical analysis of changes over time. While Landsat is one of the most comprehensive archives, it isn't the only one; for example, the global space community is bringing multimission data sets together to assess the impacts of climate change as discussed in Section 12.3. Through a combination of in situ data, modeling, and remote sensing, it has been determined that there is strong evidence that sea surface temperature has been increasing at all latitudes since the 1950s, with new satellite estimation techniques appearing to be more accurate than many in situ observations (Hartmann et al., 2013).

In theory, this wealth of remotely sensed data is a fantastic resource. However, in practice, it may not always be as simple as it sounds as often the technology onboard satellite missions has developed over time and so the data acquired now don't exactly match what were acquired historically.

In addition, remote sensing offers benefits as collecting in situ measurements is difficult to do over a wide range of areas and is costly in terms of resources; while airborne data collection could deal with the area issue, it is still expensive to fly such missions. Therefore, satellite remote sensing offers an ongoing, relatively cost-effective solution that can provide regular data acquisitions across wide geographical areas.

1.5 Challenges of Remote Sensing

Although there's a significant amount of satellite data available, and much of them are free, they are still not widely used outside the remote sensing community itself. There are a number of reasons for this: first, the sensors on satellites vary widely and, therefore, a level of knowledge is required to ensure that the data being used are the best fit for how you want to use it. Second, the data often come in a relatively raw format, requiring the user to process the data to produce the images or information they require. Third, even once processed, there's a need to understand what the data show, and the data require interpretation, which is often improved with experience. For example, a white patch on an image could show a snowy area, desert, or cloud.

Finally, satellite data sets are often vast with a large number of large files that require significant amounts of computer disk space to store and power to process. Although cloud computing has made a huge difference with the availability of remote processing, the knowledge on what's available, and access to these resources, is primarily within the research and commercial remote sensing communities.

It's also worth noting that while there are a lot of free data, they don't always cover the geographical ROI nor do they have the desired spatial resolution (see Chapter 2) needed. In these cases, it's likely the data could be available from commercial satellite operators, where there will be a cost; the purchase cost for a high-resolution optical image starts from a few hundred US dollars and can go up to several thousand dollars for very high resolution imagery.

1.6 Summary and Scope of the Book

Remote sensing provides us with the capability to answer questions about what's happening and what has happened on Earth. This book aims to show that remote sensing is a resource that's available to everyone, and it will give you the skills to start getting involved in this exciting field. It provides an overview of the science behind remote sensing and its applications, underpinning skills to enable you to process and analyze some of the data sets that are freely available, and pointers to where other data sets can be downloaded.

The book is organized such that the first seven chapters provide an introduction to the theory of remote sensing and walk you through some practical examples by showing you how to download and use both example packages and the data to enable you to start doing remote sensing. The second half of the book has five chapters focusing on different application areas, starting with urban environments, then the evolution of the natural landscape, followed by the terrestrial water cycle and then coastal environments, before finishing off with atmospheric gases and pollutants. The final chapter looks toward the future of remote sensing.

This should provide you with a good starting point for working with remotely sensed data and help you begin to investigate your own questions about our planet and its environment. However, it should also be recognized that this book can only provide an introduction to the subject, and what's possible; hence, those wishing to gain deeper knowledge should move on to more specific reading as discussed in Section 13.4.

1.7 Key Terms

- Active remote sensing: The sensor (or platform) emits energy toward the Earth, and then the sensor detects and measures the strength and time delay of the return signal.
- Earth observation: It applies to global or regional remote sensing focused on the Earth.
- Passive remote sensing: A sensor detects radiation naturally emitted or reflected by an object or region on the Earth.
- Reflectance: When light is scattered back to a sensor, with the value being the ratio of the backscattered to forward traveling light.
- Remote sensing: The collection of information about an object without being in physical contact with it.

References

Hartmann, D. L., A. M. G. Klein Tank, M. Rusticucci et al. 2013. Observations: Atmosphere and surface. In *Climate Change 2013: The Physical Science Basis. Contribution of Working Group I to the Fifth Assessment Report of the Intergovernmental Panel on Climate Change*, eds. T. F. Stocker, D. Qin, G.-K. Plattner, M. Tignor, S. K. Allen, J. Boschung, A. Nauels, Y. Xia, V. Bex and P. M. Midgley, 159–254. Cambridge: Cambridge University Press.

Lowman, P. D. 1969. *Apollo 9 Multispectral Photography: Geologic Analysis*. Greenbelt: Goddard Space Flight Center.

Sabins, F. F. 1978. *Remote Sensing Principles and Interpretation*. San Francisco: Freeman.

2

How Does Remote Sensing Work?

This chapter examines how remote sensing works, starting with the principles, followed by an explanation of what the sensors measure and how this links to the electromagnetic (EM) spectrum. It concludes by reviewing spatial, spectral, and temporal resolution. Although Chapter 1 highlighted that remote sensing can be undertaken in many ways, in this book, we're going to focus on satellite remote sensing as this is the most common type of data available to everyone.

2.1 Principles of Satellite Remote Sensing

Satellite remote sensing is primarily undertaken by sensors onboard satellites that orbit around the Earth, although there have also been instruments onboard the International Space Station. The satellites are launched into different orbits, depending on their purpose, and the two main types of orbit relevant to remote sensing are as follows:

- Geostationary orbits (a type of geosynchronous orbit): Satellites in geosynchronous orbits are approximately 36,000 km above the Earth and are the highest satellite missions used for remote sensing of the Earth itself. These satellites orbit around the equator, and their speed matches the rotation of the Earth, giving the appearance that they're stationary above a single point over the planet, giving them their name. The meteorologically oriented Geostationary Operational Environmental Satellite (GOES) system constellation is an example of a series of satellites in this type of orbit. The satellites scan 16-km² sections that are built up to form a view of the Earth, called the full Earth disk; Figure 2.1a is an example from GOES EAST acquired on December 28, 2014. The image shows the Americas (including Canada, the United States, and South America) down the center with the land being dark red and the ocean being dark blue; there's also a faint outline showing both the country borders and states of the United States. Both the land and ocean are covered in swirls of clouds that are white or pink in color, depending on their temperature and reflectance.

DOI: 10.1201/9781003272274-2

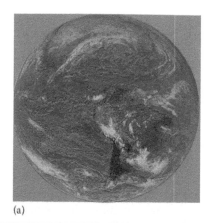

(a)

(b)

FIGURE 2.1
December 28, 2014 data. (a) GOES EAST full earth disk at 17:45 UTC and (b) all the orbits of
MODIS-Aqua. (Data courtesy of NOAA and USGS/NASA.)

- Low-Earth orbits (LEOs): These satellites orbit closer to the Earth
  than geostationary missions, at altitudes between 160 and 2,000
  km. There are two common types of LEOs, polar LEOs and sun-
  synchronous LEOs. These can be either an ascending (traveling
  south to north) or a descending (traveling north to south) orbit
  and pass close to the poles collecting data as vertical strips of the
  Earth on each successive individual orbit. Sun-synchronous polar
  LEOs are synchronized with the sun, meaning that the satellite
  visits the same point on the Earth at the same local time on each
  orbit; this is particularly useful to monitor changes over time. This
  orbit is key for small CubeSats and with the large constellations
  being launched, there are thousands of satellites in these types
  of orbits. Aqua and Terra are a pair of NASA satellites in opposite polar orbits at around 705 km, with Terra passing from north

to south across the equator in the morning, while Aqua passes south to north in the afternoon. Figure 2.1b shows an example of Moderate Resolution Imaging Spectroradiometer (MODIS) data collected on December 28, 2014, from several polar orbits of the Aqua satellite. It shows the pigment chlorophyll-a found in phytoplankton (discussed further in Section 11.1) using a rainbow color palette (with values increasing as the colors go from purple to blue to yellow to orange and then red); as this is a marine-focused product, there are no data over the land, although data will be present for lakes. The gaps in the data over the oceans are primarily attributed to clouds, which prevent optical and thermal infrared (IR) sensors collecting data.

2.2 What Does the Sensor Measure in Remote Sensing?

As noted in Chapter 1, remote sensing works by measuring the energy being reflected or emitted by an object or region of interest. Passive remote sensing uses the energy from the sun, which is emitted toward the Earth's surface (called solar irradiance), strikes the land or the ocean, and then is absorbed or scattered. The sensors ultimately measure the amount of that EM energy that's then scattered or re-emitted back up toward the detecting sensor, which is called radiance because it's captured for a narrow field of view (see Figure 2.2). The ratio of the EM energy coming down from the sun to the EM energy going back to the detecting sensor is called reflectance.

The level of reflectance and the different wavelengths involved vary for different substances on the Earth, and, therefore, each substance has its own spectral signature. Examples shown in Figure 2.3 are taken from the United States Geological Survey (USGS) spectral library (Clark et al., 2007) and Advanced Spaceborne Thermal Emission and Reflection Radiometer

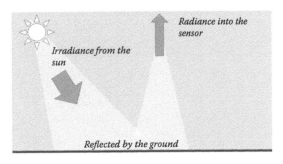

FIGURE 2.2
Process of solar irradiance being reflected by the ground and received by the sensor.

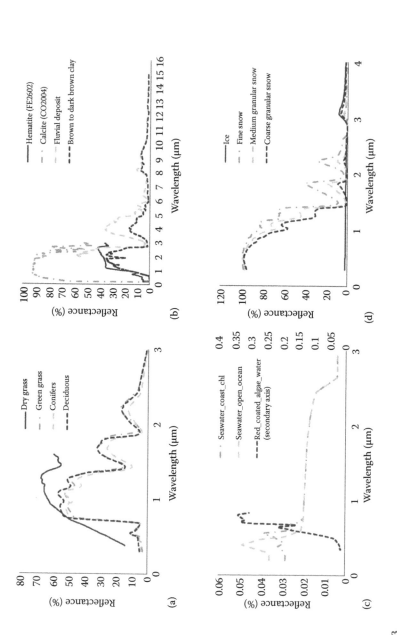

FIGURE 2.3
Spectra for different (a) vegetation types, (b) minerals that include clay soil, (c) water types, and (d) snow and ice. (Data from Clark et al, 2007, USGS digital spectral library splib06a. U.S. Geological Survey, Digital Data Series 231. http://speclab.cr.usgs.gov/spectral-lib.html; and Baldridge et al., 2009, The ASTER spectral library version 2.0. *Remote Sens Environ* 113:711–715.)

spectral library (Baldridge et al., 2009). The spectral signatures are divided into four groups that include (a) vegetation types, (b) minerals that include clay soil, (c) water types, and (d) snow and ice. The *x*-axis is variable as surfaces, such as water, don't have a strong reflectance signal beyond the visible and near-infrared (NIR), while minerals have pronounced features within the IR. The *y*-axis varies with the strength of the reflected signal; a reflectance of 100% indicates that all the radiance has been reflected back to the sensor.

Overall, the different substances and surfaces that the EM energy interacts with will determine the strength, direction, and spectral shape of the signal received by the sensor. It's worth noting that some of the energy will also be absorbed, or scattered, by the atmosphere itself as the signal passes through, both on the way to the Earth and returning to the sensor.

Active remote sensing works in a similar manner, with the exception that initially the energy is transmitted by the sensor or platform toward the Earth, and it is the absorption, scattering, and re-emission of this signal, rather than that which is coming from the sun, which is measured. In reality, it's only possible to remove the signal from the sun if the sensors are operating at night, but this solar signal will often be much weaker than the signal produced by the instrument itself.

Therefore, from knowing how a substance or surface on the Earth interacts with EM radiation, it's possible to infer what it is, although we still need to account for the effects of the atmosphere to do this accurately, a process known as atmospheric correction.

2.3 Electromagnetic Spectrum

EM radiation is energy that moves through space and matter in the form of both waves and a stream of particles called photons, with the range of EM radiation making up the EM spectrum. Multiple parts of the EM spectrum can be used for remote sensing, and each part offers the potential to interpret something different. Figure 2.4 illustrates the EM spectrum overlain with information about the radiation's wavelength. On the far right-hand side (beyond the extent of the figure) are the short wavelengths of gamma rays, less than 10^{-12} m in length, that were shown in Figure 1.1, and then the wavelengths get progressively longer until they reach the very-high-frequency (VHF) and high-frequency (HF) radio signals where the wavelength is greater than 1 m.

Two of the most relevant parts of the EM spectrum for remote sensing are the visible and IR parts. The visible part is labeled in Figure 2.4 and is often called light; the wavelengths in this region are measured in units of nanometers. Blue light is roughly situated from 400 to 500 nm; green light

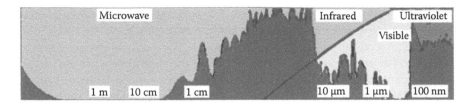

FIGURE 2.4

Electromagnetic spectrum shaded gray according to the amount of passive solar irradiance reaching the ground. (Based on a web-based figure created by European Space Agency.)

is situated between 500 and 600 nm; and red light is situated between 600 and 700 nm. As this is the part of the EM spectrum that's used for human vision, it's the most familiar to us.

As we move to the longer wavelengths in the IR part of the spectrum, which is not visible to the human eye, the wavelengths are expressed in units of micrometers. This part of the spectrum also has shorter frequencies than the visible wavelengths because there is an inverse relationship between wavelength and frequency, where the longer the wavelength, the shorter the frequency is and vice versa.

Within the IR part of the EM spectrum, the first region is the NIR with wavelengths greater than 0.7 µm, followed by shortwave IR, which can be classed as being from 0.9 to 1.7 µm, and then thermal IR, which is from 4 to 14 µm for the remote sensing of temperature.

Beyond the IR region are the submillimeter and millimeter waves that don't have a strong remote sensing focus because there are low emissions of these types of wavelengths by the sun and they are significantly absorbed by the atmosphere. In Figure 2.4, the spectral variation of the solar irradiance at ground level, which is affected by absorption within the Earth's atmosphere, is shown by the shades of gray with a lighter shade indicating less absorption; absorption is primarily by the atmospheric gases, including carbon dioxide and oxygen, plus water vapor. The lightest gray-shaded area is where there is low absorption by the atmosphere and high emission by the sun, which is particularly true for the visible wavelengths. For the shorter ultraviolet (UV) wavelengths, the atmospheric absorption rises rapidly because of ozone and thus optical missions have tended not to have wavebands within the UV. Also, the IR has significant absorption and so the effected wavelengths are, where possible, avoided.

The next part of the EM spectrum is the microwave region, which is described in terms of frequencies rather than wavelengths and runs from 40 GHz down to 0.2 GHz. Remote sensing uses specific frequency bands within the microwave region, and the main bands, starting with the longest frequencies, are as follows: Ku (12–18 GHz), X (8–12 GHz), C (4–8 GHz), S (2–4 GHz), L (1–2 GHz), and P (0.25–0.5 GHz, the shortest frequency). The instruments that use the different frequency bands will often have the band letter mentioned when they're described because different frequencies are optimal for different applications, for example, the C-band Synthetic Aperture Radar (SAR) instrument onboard the Copernicus Sentinel-1 satellites, L-band instrument on the ALOS satellites or the S-band SAR instrument on NovaSAR-1. The C-band is the most commonly used band for remote sensing, followed by the L-band and S-band with a range of applications both on the land and in the ocean. The advantage of microwave satellites is they are capable of detailed, all-weather, day-and-night observations unaffected by clouds.

The VHF and HF ranges of the spectrum are not used for satellite remote sensing, owing to their large wavelengths, but they are used within ground-based systems for detecting weather features such as rainfall and scattering off the sea surface that shows up oceanic surface currents.

2.4 How Do Sensors Take Measurements?

Sensors take measurements in a number of different ways, as described in more detail in Chapter 5, including the following:

- Photographs (originally analog and are now primarily digital) where the camera is often positioned at a nadir angle (i.e., vertically looking down) with the data captured into a two-dimensional array defined by rows and columns.
- Line scanning where the Earth is scanned line by line to gradually build up an image, using either the pushbroom or whiskbroom techniques. These are described in more detail in Section 5.1.
- Points: A number of instruments collect individual values that can later be combined during the processing to create images.

2.5 Spatial, Spectral, and Temporal Resolutions

When using satellite images, there are three important types of characteristics to understand, detailed in the following sections.

2.5.1 Spatial Resolution of Data

Spatial resolution is the size of the smallest object that can be seen on the Earth; therefore, when viewing an image with a spatial resolution of 1 km, it will not be possible to see anything smaller than 1 km, and objects would need to be significantly larger for any details to be discernible. The exception would be a very bright feature, such as a gas flare, where a small object can dominate the value of a much larger pixel. However, the size of the pixel is only one component influencing spatial resolution; for example, very high resolution features are blurred by scattering in the atmosphere, which means that it's not possible to see 0.3-m features in an image with this resolution. Also, when the term "pixel size" is used it is not necessarily referencing the resolution the sensor can image at, but rather the size of the pixel in the products you'll be given access to.

As examples, both the MEdium Resolution Imaging Spectrometer (MERIS) onboard Envisat and the sensors that are on the Landsat missions could be classed as having a medium spatial resolution; that is, their pixels have a size between 1 km and 30 m. Figure 2.5a shows an image from MERIS of Northern Europe taken on July 16, 2006, shown as the full orbit of data, while Figure 2.5b is the same image zoomed in to show the area of the port of Rotterdam, and if we keep zooming in, we eventually will be shown the actual pixels by the software, as in Figure 2.5c; in this case, for MERIS full-resolution data, the pixels are approximately 300 m × 300 m

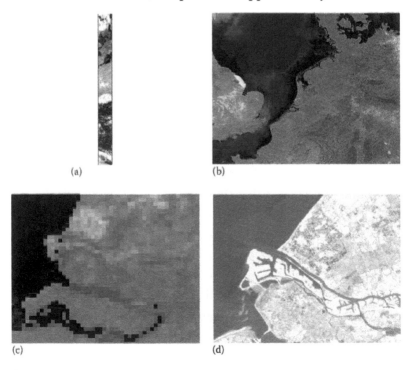

(a)   (b)

(c)   (d)

FIGURE 2.5
July 16, 2006 data. (a) An orbit of full-resolution MERIS data. (b) Zoomed-in data to see the area of interest. (c) Further zoomed-in data to see the actual pixels. (d) Higher-resolution Landsat image with a pixel size of 30 m. (Data courtesy of European Space Agency and the USGS/NASA/GLFC.)

in size. All three images are pseudo-true-color composites; Figure 2.5d shows the same area, on the same date, but with a much higher spatial resolution; the image is from the Landsat-5 Thematic Mapper sensor that has a pixel size of 30 m for its multispectral data, enabling a much greater amount of detail to be visible.

High-resolution imagery can be classed as having a pixel size of less than 30 m, and Sentinel-2 data come within this category as they have a pixel size of 10 m; while very high resolution imagery can go down to a pixel size of less than 1 m. Currently, the commercial Planet Labs satellites have a pixel size of 3.7 m, Maxar's Worldview satellites have a pixel size of just over 1 m, and their QuickBird-2 satellite has a pixel size of less than 1 m.

Figure 2.6a is a Centre National d'Etudes Spatiales (CNES) Satellites Pour l'Observation de la Terre (SPOT) High-Resolution Visible and Infrared sensor (HRVIR) image from July 16, 2009, of Grand Cayman in the Caribbean with a pixel size of 20 m; because of the wavebands being shown together, the land appears white/blue/green, and the shallow water is red. As another example, Figure 2.6b is an image taken over Bangladesh on December 14, 2004, coving the Sundarbans Reserved Forest where we can easily see individual houses and trees when it is zoomed in (inset image). The image is from QuickBird-2 and has a pixel size of 0.61–0.72 m.

Although the displayed images will have square pixels, as seen in Figure 2.5c, this is a reflection on how we handle and display data rather than the shape of the actual area captured by the sensor. The area captured within a pixel is not really rectangular but more often an ellipse shape.

Selecting the correct spatial resolution is critical to ensuring you can see what you're looking for. For example, a 1-km spatial resolution image might be fine if you're looking at changes in temperature across the Atlantic Ocean, but it won't be very useful for looking at sediment suspended in the water at the mouth of a small river.

2.5.2 Spectral Resolution of Data

Spectral resolution refers to the number of wavebands within the EM spectrum that an optical sensor is taking measurements over. There are three main types of spectral resolution:

- Panchromatic data are a single broad spectral waveband designed to capture the full visible range, but they can also be a narrower single band; for example, Landsat Enhanced Thematic Mapper Plus has a waveband between 520 and 900 nm while QuickBird-2 has a waveband between 760 and 850 nm.
- Multispectral data are collected using several wavebands at the same time, which can be combined to create color composites (see Section 5.6). Examples include MODIS and the Copernicus Sentinel-2 and Sentinel-3 missions.

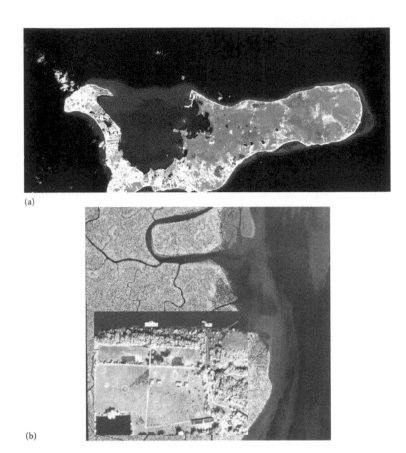

FIGURE 2.6
(a) July 16, 2009, SPOT image of Grand Cayman and (b) December 14, 2004, QuickBird-2 image of Bangladesh. (Data courtesy of ESA/CNES and GLCF/DigitalGlobe [2004], QuickBird scene 000000185940_01_P010, Level Standard 2A, DigitalGlobe, Longmont, Colorado, December 14, 2004.)

- Hyperspectral data have hundreds of wavebands and allow much more spectral detail to be seen in the image. The NASA EO-1 satellite, launched in November 2000, carried Hyperion, which was an experimental hyperspectral mission with 220 wavebands. It has since been followed by satellites such as the Italian Space Agency's (ASI) PRISMA (Hyperspectral Precursor and Application Mission), launched in March 2019, and German Space Agency's (DLR) EnMAP (Environmental Monitoring and Analysis Program) launched in April 2022.

For active microwave imagery, spectral wavebands could be replaced by the number of polarizations and viewing angles as discussed further in Section 3.2.

2.5.3 Temporal Resolution of Data

Temporal resolution, also referred to as the revisit cycle, is defined as the amount of time it takes for a satellite to return to collect data from exactly the same location on the Earth. It's expressed as a function of time in hours or days, and it depends on the satellite's orbit and the sensor swath width.

A satellite in a polar orbit takes approximately 100 minutes to circle the Earth, but with each orbit, the Earth rotates 25° around its polar axis, and so on each successive orbit, the ground track moves to the west, meaning it takes a couple of weeks to fully circle the Earth. For example, Figure 2.1b shows the swath width of MODIS as it acquires vertical stripes of the Earth.

Swath width refers to the width of the ground that the satellite collects data from each orbit. The wider the swath, the greater the ground coverage is; for example, MODIS can image the globe every 1 to 2 days as it has a swath width of 2,330 km, while Landsat-8 Operational Land Imager only has a swath width of 185 km and so the Landsat satellites have a 16-day absolute revisit time. Therefore, with a wide swath, a satellite can see more of the Earth on each orbit, and so this reduces the revisit time. In comparison, because the Sentinel data operate in pairs (see Section 6.14) their revisit time is 5–7 days, with the faster time for locations close to the equator.

2.5.4 Resolution Compromises

Unfortunately, it's not always easy to get imagery with the exact spatial, spectral, and temporal resolution needed. There is often a trade-off to be made between the different resolutions, owing to both technical constraints such as onboard memory and storage space, and the transmission capabilities from the satellite to Earth. Satellites with higher spectral resolutions generally have a lower spatial resolution and vice versa. For example, a panchromatic waveband will have a higher spatial resolution than multispectral data from the same satellite. Also, hyperspectral instruments often image only smaller areas, that is, test sites, rather than creating a complete global map.

The Sentinel missions, and previously Envisat, transmit increased volumes of data back to Earth as the European Space Agency has launched geostationary data relay satellites into the European Data Relay Satellite System (https://artes.esa.int/european-data-relay-satellite-system-edrs-overview). In addition, satellite constellations, rather than single large missions, are increasingly being launched as this increases the temporal resolution and can also be used to collect several high spatial resolution images that are joined together to form a single (more extensive geographical coverage) image. The commercial company Planet Labs has a constellation of over 200 small satellites in orbit, which can image every land mass on the Earth every day.

2.6 Summary

This chapter has focused on the theoretical side of remote sensing, looking at how the sensors acquire data. It has introduced you to the EM spectrum and the resolution triplets – spatial, spectral, and temporal – all of which you'll become much more familiar with.

We've tried to make this as simple as possible, at this stage, and each of the areas will be developed and built upon in the coming chapters.

2.7 Key Terms

- Atmospheric absorption: The absorption of EM radiation by the atmosphere as the radiation passes through it.
- Atmospheric correction: A process applied to remotely sensed data such that the absorption and scattering effects caused by the atmosphere are removed.
- Constellation: These are large numbers of satellites working together as a single system.
- Solar irradiance: The EM radiation emitted by the sun.
- Spatial resolution: The size of the smallest object that can be seen in an image.
- Spectral resolution: The number of spectral wavebands a sensor has.
- Swath width: The ground width of an image collected by a sensor.
- Temporal resolution: The frequency of data collection over a specific point on the Earth.

References

Baldridge, A. M., S. J. Hook, C. I. Grove and G. Rivera. 2009. The ASTER spectral library version 2.0. *Remote Sens Environ* 113:711–715.

Clark, R. N., G. A. Swayze, R. Wise et al. 2007. USGS digital spectral library splib06a. U.S. Geological Survey, Digital Data Series 231. Available at http://speclab.cr.usgs.gov/spectral-lib.html (accessed April 17, 2015).

3

Data Available from Remote Sensing

This chapter will describe the different types of remote sensing data available and locations on the Internet where remotely sensed data sets can be found. The satellite instruments and missions that collect the data are often referred to by their instrument acronyms, such as Moderate Resolution Imaging Spectroradiometer (MODIS) and Medium Resolution Imaging Spectrometer (MERIS), or satellite acronyms such as the Soil Moisture and Ocean Salinity (SMOS) mission. In this chapter, we will give abbreviated notations when introducing each data set.

3.1 Optical Data

3.1.1 Passive: Visible and Infrared

Passive optical data sets are probably the easiest to understand as these data include visible wavebands; therefore, the imagery can be similar to how the human eye sees the world. However, they also use wavebands beyond human vision to detect signatures related to the temperature of the Earth.

Optical instruments use a variety of spectral wavebands, but the basic distinction tends to be based on spatial resolution, particularly between medium and high-/very high resolution imagery. Examples of medium spatial resolution instruments include the following:

- Landsat series, launched from 1972 onward, with a spatial resolution of between 15 and 83 m for up to 11 spectral bands depending on the mission and the mode. The current active versions are Landsat's -7, -8, and -9; a detailed breakdown of the series can be seen in Table 6.1.
- MERIS, launched in 2002 and ceased producing data in 2012 when the Envisat satellite it was onboard stopped functioning. MERIS had spatial resolutions of 300 and 1,200 m for 15 spectral bands and was the forerunner to the Ocean and Land Colour Instrument (OLCI) onboard the Sentinel-3 missions.

DOI: 10.1201/9781003272274-3

- MODIS, launched on separate satellites in 1999 and 2002, with a spatial resolution of between 250 and 1,000 m for 36 spectral bands.
- Visible Infrared Imaging Radiometer Suite (VIIRS), launched in 2011 on board the joint the National Aeronautics and Space Administration (NASA)/National Oceanic and Atmospheric Administration (NOAA) Suomi National Polar-orbiting Partnership (Suomi NPP) and NOAA-20 satellites, with a spatial resolution range of between 375 and 750 m for 22 spectral bands.
- Sentinel-2 constellation of two satellites, 2A and 2B, launched in 2015 and 2017, respectively, with further missions named 2C and 2D planned to be launched. The MultiSpectral Instrument (MSI) has a spatial resolution of 10 m and 13 spectral bands.

The following are examples of high- to very high resolution instruments:

- Satellites Pour l'Observation de la Terre (SPOT or Earth-observing satellites) series, began in 1986 with the launch of SPOT-1. At the point of writing, SPOT-6 and SPOT-7 are the operational satellites, launched in September 2012 and June 2014, respectively, which operate in tandem with 6-m pixels for four spectral bands and panchromatic data at 1.5 m for a 60 km × 60 km swath.
- Pleiades 1A, launched in December 2011, followed by 1B, launched in December 2012, have 0.5-m pixels across four spectral bands primarily aimed at the commercial marketplace.
- WorldView-3 and WorldView-4, launched in 2014 and 2016, respectively, offering 0.31-m pixels for their panchromatic band and 1.24 m for their 16 spectral wavebands and 3.70 m for their shortwave infrared wavebands. It also included the Clouds, Aerosols, Water Vapor, Ice, and Snow (CAVIS) instrument that aids with atmospheric correction.
- Planet Labs operates satellite constellations, with approximately 200 satellites in orbit at any one time – although this number fluctuates as new launches occur and the older satellites burn up on re-entry. These satellites have a range of spatial resolutions of between 5 and 0.5 m, and between them, Planet Labs reports that they image every landmass on Earth daily.

As highlighted in Chapter 2, the spectral and spatial resolution varies between, and within, satellite instruments. Therefore, as shown for the listed missions, not all spectral bands will have the highest available spatial resolution.

3.1.2 Active: Lidar

Lidar, which stands for Light Detection and Ranging, is an active remote sensing technique that uses a laser scanner to map the Earth's topography by emitting a laser pulse and then receiving the backscattered signal. There are two main types of Lidar: topographic and bathymetric; topographic Lidar uses a near-infrared laser to map land, while bathymetric Lidar uses water-penetrating green light to measure the depth of the seafloor.

Airborne laser scanning is most common because of the power requirements placed on a satellite mission by an active system. However, there have been satellite Lidars such as the following:

- The Cloud-Aerosol Lidar and Infrared Pathfinder Satellite Observation (CALIPSO), a joint Centre National d'Etudes Spatiales (CNES)-NASA mission launched in 2006, aimed to provide new insights into the role that clouds and atmospheric aerosols (airborne particles) play in regulating Earth's weather, climate, and air quality. However, this ongoing satellite, with a long time series of data that is far beyond its 3-year designed life span, is providing further insights such as profiles of backscatter from particles in the ocean.

- Ice, Cloud, and Land Elevation Satellite-Geoscience Laser Altimeter System (ICESat-GLAS) operated between 2003 and 2009, and then ICESat-2 was launched in September 2018. Their primary objective has been measuring not only the ice sheets and glaciers but also the height of vegetation such as trees.

- ADM-Aeolus satellite, launched in 2018 by the European Space Agency (ESA), carries an Atmospheric Laser Doppler Instrument, which is essentially a Lidar instrument that provides global measurements of wind profiles from the ground up to the stratosphere with 0.5–2 km vertical resolution.

3.2 Microwave Data

A second type of remote sensing data is from the microwave part of the EM spectrum and has an advantage over optical data in that the signal can penetrate clouds and smoke, and sometimes precipitation; hence, it is not so strongly affected by weather conditions. Several types of microwave instruments use the time of travel to measure heights and determine different surfaces from their roughness or dielectric constant, which is the extent to which a material concentrates an electric flux.

3.2.1 Passive: Radiometer

Passive microwave sensors, often called microwave radiometers, detect the natural microwave radiation emitted from the Earth. The spatial resolution is relatively coarse, often tens of kilometers, and this weak signal is prone to interference from external noise sources such as microwave transmitters. Radiometers measure soil moisture, ocean salinity, sea ice concentrations, and land and ocean surface temperature.

As an example, the ESA SMOS mission, launched in November 2009, carries a radiometer, and as its name suggests, it's a soil moisture and salinity-focused mission, although it has also been used for several other applications, as happens for many instruments.

3.2.2 Active: Scatterometer

Scatterometers send out pulses of microwaves in several directions and record the magnitude of the signals scattered back to the sensor. They're commonly used to measure wind speed and direction, used mainly for operational meteorology. For example, the NASA SeaWinds instrument onboard Quick Scatterometer (QuikSCAT) measured winds over the ice-free ocean daily from July 1999 to October 2018. Subsequent missions include the Indian Space Research Organization (ISRO) Oceansat-2 scatterometer, known as OSCAT, launched in September 2009, and the ASCAT instrument on the European Organisation for the Exploitation of Meteorological Satellites (EUMETSAT) MetOp satellites.

3.2.3 Active: Altimeter

This is an active sensor used to calculate heights on the Earth by sending out short bursts, or pulses, of microwave energy in the direction of interest with the strength and origins of the received back echoes or reflections measured. The signal is corrected for several potential error sources, for example, the speed of travel through the atmosphere and small changes in the orbit of the satellite. Applications include calculating the height of the land, ocean, and inland water bodies. Examples include the following:

- Jason-2 Ocean Surface Topography Mission, launched in 2008, carries the Poseidon-3 Radar altimeter, Advanced Microwave Radiometer (which allows the altimeter to be corrected for water vapor in the atmosphere), and instruments to allow the satellite's position in space to be determined very accurately. Jason-2 followed Jason-1, launched in 2001, and is preceded by Jason-3, launched in 2016.

- Sentinel-3A and 3B launched in 2016 and 2018, carrying a dual-frequency Synthetic Aperture Radar (SAR) altimeter called the SAR Radar ALtimeter (SRAL) instrument, among several other instruments.
- Copernicus Sentinel-6 Michael Freilich was launched in November 2020, with Sentinel-6B due for launch in 2025. It is a continuation of the abovementioned altimetry missions and will take over from Jason-3 as the reference altimetry mission. The name "Michael Freilich" is in honor of the former Director of NASA's Earth Science Division.

3.2.4 Active: Synthetic Aperture Radar

SAR is an active sensor where the pixel brightness is related to the strength of the return signal, with a rougher surface producing a stronger radar return and the resulting pixels appearing brighter in the imagery. It's called SAR as it uses a small physical antenna to imitate a large antenna because detecting long microwave frequencies in space would require a physical antenna thousands of meters in length; however, the same result can be achieved with a synthetic antenna of approximately 10 m in length. This is possible because as the satellite moves, all the recorded reflections for a particular area are processed together as if they were collected by a single large physical antenna.

The orientation of microwave signals is known as polarization. Signals emitted and received in horizontal polarization are known as HH signals, and those emitted and received in vertical polarization are known as VV signals. Cross-polarized signals, such as HV or VH, are also possible.

SAR data are influenced by the azimuth, the direction the sensor is looking, and the orientation of the object of interest. Smooth surfaces have a mirror-like reflection, where the sensor only measures a return signal when it is directly above the target, and these surfaces appear dark in SAR imagery as the sensors are normally measuring at an angle. However, large, flat, rectangular surfaces, such as building facades that extend upward, are oriented perpendicular to the signal and, thus, can act like corner reflectors enhancing the like-polarized return, meaning that a horizontally polarized signal sent out will have an enhanced horizontally polarized return signal. Therefore, the same area can appear differently on different SAR images depending on the choice of polarization and look-angle.

Historically, there have been limited SAR satellites launched; however, this has changed over the last decade with several significant launches, including the following:

- Sentinel-1A and 1B launched in 2014 and 2016, respectively, and orbited 180° apart; although Sentinel-1B ceased operations in January 2021, but there are plans to launch Sentinel-1C in 2023, and Sentinel-1D will follow this. They carry a C-Band SAR instrument and operate in several modes, the most common being the Interferometric Wide swath mode where each scan has a width of 250 km and a spatial resolution of 5 m by 20 m.

- ICEYE is a Finnish company, and they have launched over 20 SAR satellites since 2018, operating as a constellation, with the intention that this number will increase. Each satellite has an X-Band SAR instrument operating in VV polarization, with a spatial resolution down to 3 m with a scene size of 30 km by 50 km.

- Capella Space, a US firm, has a small constellation of X-Band SAR satellites, offering a very high resolution of spatial resolution between 0.5 and 1.2 m depending on mode, with scene sizes ranging from 5 × 5 km to 5 × 20 km.

- A specially modified SAR system with C and X bands was flown onboard the space shuttle *Endeavor*, during an 11-day mission in February 2000. This produced the Shuttle Radar Topography Mission (SRTM) Digital Elevation Model (DEM) global data set (https://www2.jpl.nasa.gov/srtm/) that, as of 2014, began to be made available globally at 30 m resolution. This is of particular interest to remote sensors as it allows topographical features to be understood; for example, we'll use these data to add contour lines to imagery in Chapter 10.

3.3 Radio Data

The Global Navigation Satellite System (GNSS), including GPS, are signals initially designed to measure position with your mobile phone having a sensor to detect the signal and work out where you are. However, these signals can also be used to determine other important data such as temperature, pressure and humidity profiles through the Earth's atmosphere for applications such as weather forecasting. Using the data in this way is called radio occultation (RO).

Spire's global constellation is composed of LEMUR nanosatellites are equipped with different sensors to collect global radio frequency data in near-real time. The satellites have RO capabilities and the ability to receive signals sent from ships at sea, providing worldwide tracking of maritime operations. Similarly, aircraft broadcast signals allow them to be tracked globally.

The value of RO data was increasingly recognized, and so this capability was added to the Copernicus Sentinel-6 Michael Freilich mission. Measurements of RO are also made by EUMETSAT's MetOp-Second Generation satellites, with the primary objective being to maintain the EUMETSAT Polar System's role as the source of observations with the highest positive impact on global numerical weather prediction up to 10 days ahead.

3.4 Distinction between Freely Available Data and Commercial Data

Another distinction that can be made is between the data that are freely available to the general public and the commercial data that need to be paid for. Historically, space agencies launched global missions and have produced large data sets with medium to low spatial resolutions, and the majority of these data sets are made freely available in various formats.

The most well-known space agency is probably NASA, and they've had a long program of launching satellites to observe the Earth. Other space agency examples include ESA, the China National Space Administration (CNSA), the Indian Space Research Organization (ISRO), and the Japan Aerospace Exploration Agency (JAXA); the full list is available from Committee on Earth Observation Satellites (CEOS, https://ceos.org/about-ceos/agencies/).

In addition, national and international organizations work with space agencies to deliver operational services; for example, NOAA provides weather forecasting and weather warnings in the United States. Again, the data from these organizations are often freely available. This has been taken one step further in Europe, with the European Commission running the Copernicus program of Sentinel missions. ESA, EUMETSAT and ECMWF are responsible for coordinating the data products and services provided by Copernicus, which are all freely available to any user.

Therefore, with lots of medium- to low-resolution data freely available, commercial organizations have focused on providing high- and very high resolution imagery. These data have applications ranging from military intelligence to security and mapping, which was historically undertaken through an airborne survey.

Very high resolution imagery will result in a large computer file (see Section 5.3), and there are constraints on the storage and transmission of files for satellites carrying these instruments, which means that commercial satellites don't always routinely collect global data – although this is changing with recent commercial CubeSat operators. Instead, when data are required, satellites are individually pointed toward the region of interest, a process known as tasking. This tasking generates a cost for the data collection, which is why commercial organizations charge extra for it. The cost can be lower if a previous user has already requested the data, and so the data are already sitting within an archive. Alternately, several space agencies are now bulk buying commercial data and providing it for scientific use rather than launching their own satellites to collect such data.

Costs vary among commercial operators, but generally, the higher the spatial resolution, the higher the data purchase price. However, it can also be cheaper to purchase higher-resolution data as there will be a minimum area that needs to be purchased – so if you're just interested in a very small area such as a field look at the different process before making a choice.

Operators of commercial satellites include the following:

- Maxar with their GeoEye, IKONOS, QuickBird, and WorldView high- and very high resolution sensors, down to 0.3-m pixels.
- Planet Labs with their optical CubeSat constellations containing Dove, Pelican, and SkySat satellites, plus the five RapidEye very high resolution satellites.
- Commercial SAR providers such as ICEYE and Capella Space.
- Chang Guang Satellite Technology Co. Ltd is a Chinese company that developed the Jilin commercial satellite constellation that, by the end of 2025, is planned to consist of 138 satellites providing all round-the-clock, all-weather data acquisitions.
- Spire for meteorological and ship/aircraft positioning data.
- Surrey Satellite Technology Ltd operates the UK Disaster Monitoring Constellation (DMC) missions, a series of small satellites carrying three-waveband, medium-resolution optical sensors with a wide swath of around 600 km and higher-resolution five-waveband sensors with a smaller swath, to provide daily global optical imaging of the Earth. They've also developed NovaSAR-S (an S-band Radar) that aims to deliver all-weather medium-resolution microwave data at a price similar to commercial optical missions.

There are also joint ventures between space agencies and commercial organizations to design, build, and launch remote sensing satellites. The following are examples:

- Airbus DS designed and built both Cryosat-2 and SPOT 7, and was responsible for building the Sentinel-1 SAR instrument in addition to being the lead satellite contractor for the Copernicus Sentinel-2 missions. They also collaborated with the German Space Agency, Deutsche Forshungsanstalt fur Luft and Raumfahrt (DLR), on the TerraSAR-X in June 2007 and then the TerraSAR-X add-on for Digital Elevation Measurement 2011 (Tandem-X) in June 2010.
- Raytheon designed and manufactured VIIRS.
- RADARSAT Constellation Mission is a three-strong constellation of SAR satellites built and operated by the Canadian Space Agency in partnership with several industrial partners including MDA and Northrop Grumman Astro Aerospace.
- Thales Alenia Space was the lead for designing and building the Copernicus Sentinel-1 and Sentinel-3 missions. They also collaborated with the Agenzia Spaziale Italiana (ASI) on the COSMO-SkyMed constellation, a dual-use (civil and military) program with customers including public institutions, defense organizations, and the commercial sector.

3.5 Where to Find Data?

There is currently no single location where all remote sensing data can be accessed from, but there is an increasing number of initiatives focused on creating data portals that provide either direct access or onward links to data from a range of sources for a particular application.

The historical route for obtaining the data, which remains valid, is to go to the organization responsible for capturing and/or processing the data. However, you can now also set up your processing on cloud-based platforms where the data are made available to you online, so you don't need to download it. The most famous of these is GoogleEarthEngine, which many scientists use on a free-to-access basis, but there are other options provided by Amazon Web Services and Microsoft alongside organizations such as EUMETSAT and EO-focused companies such as CloudFerro; we will discuss this more in Chapter 13.

Several continually updated web links are provided via the online resource (see Chapter 4) alongside some of the longer-term websites listed below:

- ESA Online Dissemination for Landsat data that have been received by ESA ground stations: https://landsat-diss.eo.esa.int/oads/access/

- ESA Principal Investigator Community, a scientifically focused portal for accessing data held by ESA that includes third-party mission data purchased from commercial operators: https://earth.esa.int/eogateway/activities/pi-community
- Copernicus Services: https://www.copernicus.eu/en/copernicus-services
- CREODIAS run by CloudFerro: https://creodias.eu/
- Global Earth Observation System of Systems (GEOSS) portal, which provides an interactive way to find out about how to access a large number of data sets: https://www.geoportal.org/
- Google Earth Engine: https://earthengine.google.com/
- NASA's Geospatial Interactive Online Visualization ANd aNalysis Infrastructure (Giovanni) Portal, an online tool that allows you to produce images and plots without downloading data: https://giovanni.gsfc.nasa.gov/giovanni/
- NASA OceanColor Web, which has ocean color, salinity, and sea surface temperature from a number of missions: https://oceancolor.gsfc.nasa.gov/
- NOAA Comprehensive Large Array-data Stewardship System (CLASS), for accessing all the data held by NOAA (including VIIRS data): https://www.avl.class.noaa.gov/saa/products/welcome
- USGS Landsat archive, the main route for accessing Landsat data: https://www.usgs.gov/landsat-missions/data/
- WEKEO developed by EUMETSAT, European Centre for Medium-Range Weather Forecasts (ECMWF), Mercator Ocean International, and the European Environment Agency (EEA): https://www.eumetsat.int/who-we-work/wekeo

3.6 Picking the Right Type of Data for a Particular Application

The choice of which data source to use will be a combination of the following:

- Specification requirements for the application, such as whether it needs optical or microwave data, specific spectral wavebands/frequency bands, spatial resolution, and revisit time.

- Whether the data have to be freely available or it can be purchased.
- Availability, in terms of which satellites cover the area required— not all satellites have global coverage, although there is a limited pointing capability for the high- to very high resolution missions.
- Lead time from ordering to the delivery of the data.

As we started discussing at the end of Chapter 2, the decision on which data to use involves making trade-offs:

- Revisit time: Landsat collects data with a revisit time of 16 days and provides regular, but not daily, coverage. Many applications require a revisit time of less than a week; hence, the Copernicus Sentinel-1 and Sentinel-2 missions use a pair of satellites to provide a revisit time of 5 days at the equator and 2–3 days at midlatitudes. Very high resolution satellites can be tasked to acquire data quickly and can often be pointed to acquire data from more than one possible orbit, but the faster the data are needed, the higher the cost is likely to be. Constellations, DMC, and small satellite operators like Planet Labs, have several satellites so they can try to attain daily coverage.
- Spatial resolution: In addition to the influence of spatial resolution on what is visible within an image, as the resolution decreases, the ability to identify different components tends to decrease. Therefore, relatively small pixels are needed for characterizing highly variable areas, whereas larger pixels may suffice for characterizing larger-scale variability.
- Required spectral wavebands/frequency bands: Depends on the spectral features and microwave signatures required. For different applications, different sections of the EM spectrum will be most optimum. These will be discussed, along with the approaches/algorithms used to derive quantitative results, in the applications chapters within the second part of the book.
- Cost: When considering cost, it's necessary to think about both the cost of acquiring the data and the cost of the data analysis. Images that are more highly processed are likely to cost more, although cheaper raw data will take time, and knowledge, to process. The cost price is often also related to spatial resolution, with the higher the resolution, the higher the cost. However, recent developments of smaller and lower-cost satellites aim to reduce the cost of high-resolution imagery and are discussed further in Chapter 12.

3.7 Summary

Chapter 2 introduced you to how satellite remote sensing works, whereas in this chapter, we've focused on the data collected by those satellites. Understanding the different types of data available and their different characteristics forms the basis of remote sensing, and we've also given you some signposts on where to find these different types of data.

In the second half of the book, we'll describe the most appropriate type, or types, of data to use for individual applications. The practical exercises will be mainly focused on visible and infrared data, as these are easier to access, although we'll also be using some SAR data in the later chapters.

3.8 Key Terms

- Altimeters: Active sensors used to measure heights on Earth through measuring the return time of emitted pulses.
- DEM: Three-dimensional representation of the Earth's surface in terms of elevation.
- Dielectric constant: Number that indicates the extent to which a material concentrates electric flux.
- Polarization: Property of EM waves that can oscillate with more than one orientation.
- Radiometer: Passive sensor that measures the natural EM radiation emitted or reflected from the Earth.
- SAR: Uses a small physical antenna to imitate having a large physical antenna by taking advantage of the movement of the satellite to collate multiple recorded reflections that are processed together as if they were collected by a single large physical antenna.
- Scatterometer: Active sensor that sends multiple microwave pulses in several directions and uses the returned signals to determine parameters such as wind speed and direction.

4

Basic Remote Sensing Using Landsat Data

The next four chapters will introduce the knowledge and techniques for handling, processing, and analyzing remotely sensed data. They will provide a background to image processing theory and a step-by-step guide to applying basic remote sensing techniques.

They'll also give details on obtaining free software and freely available remote sensing data. We will concentrate on using data from the Landsat and Sentinel missions as the demonstration data sets for this introduction, as they are easily accessible through the Internet and free of charge to use.

This chapter introduces Landsat and the concepts of finding, downloading, and viewing that source of remotely sensing data. In addition to this book, there's an accompanying website (https://www.playingwithrsdata/) that explores the practical activities in greater detail and provides up-to-date information as software is revised and data sets are reprocessed.

4.1 Notation Used for Practical Exercises within the Book

This is the first chapter of the book containing the remote sensing practical exercises, and to support learning, a consistent notation is used to indicate what is being referred to.

- All websites are shown in parentheses.
- Where you're asked to go through different menu layers, a > sign indicates each menu level to be selected. For example, if you need to select the File menu and then the Save option, it would be shown "go to the menu item File > Save."
- Any text you're required to enter into a website or package is enclosed within double quotes, "like this".

DOI: 10.1201/9781003272274-4

4.2 History of Landsat

The first Landsat mission was ERTS-1 (Earth Resources Technology Satellite), later renamed Landsat-1, which was launched into a sun-synchronous near polar orbit on July 23, 1972. It carried a Multispectral Scanner (MSS) that had four spectral wavebands within the green, red, near-infrared (NIR), and infrared (IR) parts of the electromagnetic spectrum and supplied data with a 60 m pixel size.

The main achievement of Landsat has been the acquisition of an archive of more than 50 years of remote sensing data, which have been delivered through the launch of subsequent Landsat satellites coupled with a desire to ensure data continuity between the different missions. The five-decade journey hasn't always been smooth. Landsat was established by the National Aeronautics and Space Administration (NASA), then transferred to the private sector under the management of the National Oceanic and Atmospheric Administration (NOAA) in the early 1980s, and, after operational challenges, returned to US Government control in 1992. There have also been technical issues, most notably the failure of Landsat-6 to reach its designated orbit and Landsat-7 suffering a partial failure in 2003 (described in more detail in Section 4.11).

Despite the behind-the-scene concerns and changes of operation, the Landsat data sets have become a successful global resource with 10 million scenes available in the archive by 2021. Usage dramatically increased in 1980 when the United States Geological Survey (USGS) decided to make the data freely available, and more than a million scenes were downloaded in the first year compared with a previous annual high of around 25,000 scenes for commercially sold imagery (Wulder and Coops, 2014). By the start of 2015, more than 22 million scenes had been downloaded through the USGS-EROS website with the rate continuing to increase (Campbel, 2015).

In addition, the European Space Agency (ESA) also provides Landsat data sets downloaded from their own data-receiving stations. However, the ESA archive is useful as the ESA data set includes data over the ocean, whereas USGS data sets do not, and so a web link is provided in Section 4.14. According to NASA, across all the available options, the 100 millionth Landsat scene was downloaded by a user in 2020.

Finally, Landsat-5 officially set a new Guinness World Record title for the "Longest-operating Earth observation satellite" with its 28 years and ten months of operation when it was decommissioned in December 2012.

4.3 Summary of the Landsat Missions

Eight Landsat missions have collected data, and at the time of writing, the current active missions are Landsat-8 and Landsat-9, with Landsat-9 launched on September 27, 2021. The next Landsat mission will be known as Landsat Next and is already planned for launch in 2029 or 2030. It is hoped that this new mission will collect data from significantly more spectral bands. While data continuity has been a key aim of Landsat, as technological capabilities have increased, the instruments onboard consecutive missions have been improved, and so there are four "families" of missions:

- Landsat missions 1, 2, and 3 had both the MSS and Return-Beam Vidicon (RBV) instruments onboard. As previously mentioned, MSS scenes are provided with images having a pixel size of 60 m – although the original pixel size is 79 by 57 m, this is resampled to provide users with square pixels that make data handling easier. In addition, the four MSS wavebands on the first two missions were supplemented by a fifth waveband for the thermal IR (TIR) on Landsat-3, designed to allow nighttime operation, but it never became operational. The RBV consisted of three cameras, one for each spectral band (waveband 1: blue-green, waveband 2: yellow-red, waveband 3: NIR) that measured reflected solar radiation with an 80-m spatial resolution, although the RBV on Landsat-1 only operated for a short period.

- Landsat missions 4 and 5 had the historical MSS instrument, to provide data continuity, and a new Thematic Mapper (TM) instrument to replace the RBV. TM increased the available wavebands from four with MSS to seven, giving a blue waveband alongside additional IR and TIR wavebands; in addition, the digitization increased from 6 to 8 bits. TM had a 30 m pixel size for the reflective wavebands and 120 m for the TIR waveband, although these data are resampled to a 30-m pixel size to make the data handling easier, but it doesn't actually improve the spatial resolution.

- Landsat-7 has an Enhanced Thematic Mapper Plus (ETM+) instrument with the same six reflective wavebands of the TM with a pixel size of 30 m, with a TIR waveband at 60 m resampled to the optical data resolution of 30 m. This instrument also has a panchromatic waveband at 15-m resolution. The digitization of the data increased again to 9 bits.

- The two satellites are operationally identical, although Landsat-9 produces a greater volume of data. Landsat-8 and Landsat-9 both carry an Operational Land Imager (OLI) and a Thermal Infrared Sensor (TIRS) onboard, although on Landsat-9 they are the Operational Land Imager 2 (OLI-2) and Thermal Infrared Sensor 2 (TIRS-2). The OLI and OLI-2 sensors have eleven wavebands, four visible spectral bands, one NIR, and three SWIR bands, all of which have a pixel size of 30 m, plus one panchromatic band at 15 m pixel size, and lastly, there are two thermal bands at 100 m pixel size. TIRS and TIRS-2 contribute the same two TIR wavebands with a spatial resolution of 100-m, resampled to a pixel size of 30 m, although TIRS-2 is designed to minimize stray light. The digitization is also improved again, to 12 with Landsat-8, then 14 bits with Landsat-9, resulting in imagery with over 16,000 gray levels.

A summary of the Landsat missions, their sensors, and spectral wavebands can be found in Table 6.1. For more information on each Landsat instrument, including the data in detail, the USGS web pages (https://www.usgs.gov/landsat-missions/landsat-satellite-missions/) provide a comprehensive resource.

As there are no onboard recorders on the Landsat missions, data acquisitions are limited to a real-time (RT) downlink only. Therefore, global data collection relies on a network of receiving stations called the International Ground Station (IGS) Network; details are available online (https://landsat.usgs.gov/igs-network/), where it lists the current and historical ground stations. We have provided links in the Online Resources (Section 4.14) to useful links for accessing and investigating Landsat data.

4.4 Different Levels of Data Available

Remotely sensed data are often described as being processed to different levels or stages. For globally collected data, there are four possible levels of processing:

- Level 0 (L0) refers to the raw data. These are the data received from the satellite simply converted into a format suitable for storing. This level of data is primarily of interest to experienced remote sensing users and, thus, is not always distributed.
- Level 1 (L1) data are L0 data that have been processed into formats that are more accessible to users, such as images. For Landsat, this

is described in more detail in Section 6.5 as this is the level primarily made available.

- Level 2 (L2) adds several derived geophysical parameters such as ocean wave height, soil moisture, or ice concentration. The L1 to L2 processing would include an atmospheric correction for optical sensors.
- Level 3 (L3) data are composite data sets where large areas, often the whole globe, are mosaicked together from multiple orbits. Figure 2.1b was an example of this level, containing a day of MODIS-Aqua data.
- Level 4 (L4) data have additional processing, such as gap filling through interpolation or combining data from several sensors. An example of this data type is multi-mission time-series data sets for climate analysis.

The Landsat archive uses Landsat Collection 2, which is a reprocessing of all the data in the archive collected by the Landsat 1 to –9 missions, from 1982 to the present day with improvements to geolocation accuracy, calibration, and validation. In addition to the top of atmosphere products originally available for Collection 1, Collection 2 contains global L2 surface reflectance and surface temperature scene-based products. Access to the Landsat Collection 1 data was discontinued at the end of 2022.

With each collection, there are different Tiers of data. RT data are newly acquired from Landsat-8 and Landsat-9 and are available in less than 12 hours (4–6 hours typically). Tiers 1 and 2 indicate the data's final quality, with the RT tier reprocessed. Tier 1 is of the highest geometric quality and is considered suitable for time-series analysis. Scenes not meeting this standard are classed as Tier 2, which could be due to less accurate orbital information (e.g., for older Landsat missions), significant cloud cover, insufficient ground control, or other factors. The Tier designation is visible at the end of the Landsat Scene Identifier.

The rest of this chapter will focus on the basic use of L1 data, although the application chapters in the book's second half will describe how some of the other data levels can be used.

4.5 Accessing the Level 1 Landsat Data

Landsat data are available from various sources; however, USGS is the primary source as Landsat is a joint venture between USGS and NASA, with USGS holding the data collected by the full IGS Network. This chapter will use the USGS Landsat archive website (https://landsat.gsfc.nasa.gov/data/where-to-get-data/) where there are three options to find and download global data easily:

- LandsatLookViewer
- Global Visualization Viewer (GloVis)
- EarthExplorer

While all three simple options offer access to Landsat data, for the purposes of this chapter, we're going to use GloVis because we believe that it's the easiest tool for new users to work with, being a browser-based viewer. We'll discuss and download data through the EarthExplorer option in later chapters. The first step for any system is registering and creating a user account. The accounts are free to create, and you will need to create a username and password, enter contact information, including an email, and answer a few questions on how you intend to use the data to help USGS assess their user base. However, once created and activated, you can download as much data as you need.

Although most common web browsers will work effectively with the Landsat USGS website, they need to be up-to-date as the website only supports current and recent versions.

4.6 Selecting the Level 1 Landsat Data to Download

After activating your account, the next step is to locate a Landsat image using the GloVis website (https://glovis.usgs.gov/). When GloVis is opened, you'll see the home screen. On it there is a button to "Launch GloVis". In the menus across the top, there is an option to take a tour of the software, and you are welcome to do this whenever you want. If you are not going to do the tour now, press the "Launch GloVis" button.

You'll see the main screen of GloVis (see Figure 4.1) has a left sidebar with the Interface Controls, the main screen with the map showing the current area of interest, and in the lower left of the main image is the Scene Navigator panel. If you can't see your Scene Navigator panel, click on the icon that looks like a steering wheel at the top of the Interface Controls panel to hide and unhide it. In the center of the main screen is a map with the target cursor – a small black circle with bars at each quarter point – in the center indicating the current position. In the information bar at the top of the main image are the latitude and longitude of the center

FIGURE 4.1
Screenshot of the USGS GloVis Browser. (Courtesy of U.S. Geological Survey.)

point. In the top right of the main screen is the base layers icon, which is three rectangles one on top of the other like three sheets of paper, which gives you options on how you see the main image. The default option is OpenStreetMap, but several other cartographic, imagery, or topographic options exist.

The first step is to select the geographic location you are interested in, which can be done in several ways:

- Move Across the Map Yourself: Put your mouse over the map in the main screen; the cursor will change to a small hand. Hold down the left mouse button, and move the mouse; the map will move in whichever direction you choose, with the target cursor staying in the center of the screen. When you let go of the left mouse button, the target cursor shows the point chosen. You can zoom in and out using either the + and – buttons in the top right-hand corner of the main screen or the mouse wheel if you have one. In a broader view, you can use the same technique described above with the small hand cursor to move around the globe.

- Jump To Location/Scene: This can be found by clicking the "Jump To – Click for options to jump to a specific location/scene" icon, which looks like a map and compass, next to the latitude and longitude coordinates at the top of the main screen. This opens the Jump To Location/Scene dialog box, where you have four options:
  - Current Location: This jumps the map's center point to your current location on the planet, based on your IP address, assuming you allowed GloVis to know this information.

- Lat/Lng: Clicking on Lat/Long allows a specific latitude and longitude to be directly entered. There are two things to note about inputting this data into GloVis. First, GloVis uses the decimal version of latitude and longitude, rather than the degrees, minutes, and seconds for easy data entry. The degrees, minutes, and seconds version can be easily converted into the decimal version using

$$Decimal = degrees + (minutes / 60) + (seconds / 3,600) \qquad (4.1)$$

Second, northern latitudes are entered as positive numbers, while southern latitudes are negative numbers, eastern longitudes are positive, and western latitudes are negative. Once the latitude and longitude have been entered, press Jump To Location button, and the main image will move to the new location.

The cursor's latitude and longitude are shown at the top of the main image.

- Scene ID: Clicking on Scene ID opens this option. Every Landsat scene has a specific ID number, described further below in Section 4.7. Select the data set, enter the scene ID and whether you want it to display the scene ID, and press Jump To Scene, it will move directly to the selected scene. Obviously, you need to know the Scene ID to use this option.

- WRS Path/Row: This refers to the Worldwide Reference System (WRS), which is a specific form of notation used to catalog Landsat data by specifying a path and row and is described in detail within Section 4.8. Clicking WRS Path/Row allows you to enter the specific Path and Row, then Press Jump To Location to move to the selected location.

Once you've got the geographic location you are interested in, the next step is to select the data set you want to view, using the "Selected Data Set(s)" input box under the Interface Controls panel on the left side. Initially, this will show "No data sets selected, add a data set to continue". Press the + symbol to give you the list of data sets available in GloVis. Use the scroll bar on the right side to go down, and you'll see all the different Landsat missions, Sentinel-2, plus several other satellite missions and other data sets. For this chapter, scroll down until you find "Landsat 8–9 OLI/TIRS C2 Level 1" and click the right-hand side of the toggle switch to turn it on. Press "Add Selected Data Sets" button at the bottom of the window. This will now overlay the base layer with a Landsat 8–9 image.

Underneath the Selected Data Set(s), it will show how many images are available for the data set and your chosen geographic location. Remember that the color image is a Landsat scene that covers an area of 185 km², and you may need to zoom back in to see it.

All the panels can be closed by simply clicking on the *x* in the panel's top right corner.

Having selected the data set and geographical location you are interested in, there are likely to be many images available, possibly hundreds. While you could scroll through the images one by one using the Scene Navigator as described below, it is more useful to filter the images to reduce them to the specific ones you are interested in. This filtering can be done using the Common Metadata Filters dialog box under the Interface Controls. If you click on the + sign next to Common Metadata Filters you'll open a new window with a Filter Type box, click on the down arrow at the end of the box to drop down, and you'll have the following filter options:

- Acquisition Date: To enter a specific time period, use a start and end date entered using dd/mm/yyyy.
- Cloud Cover: As Landsat is an optical sensor, it cannot see through clouds, and therefore, heavy cloud cover may prevent the region of interest from being seen. This option allows you to choose the percentage of the scene you're willing to have covered by cloud. By selecting 0–100, GloVis may offer images totally obscured by clouds, whereas selecting 0–40 will only give images with 40% cloud coverage or less. There is also a slider as an alternative way of choosing the Cloud Cover values.
- Seasonal (Month Selections): To enter a specific period of the year. To select more than one month, hold down the Shift or Ctrl key while making your choice.
- Scene Ingest Date: To enter a specific time period for when the scene was ingested into the system using two dates entered as dd/mm/yyyy.

After choosing the Filter you want, press the Add Filter button and the Close button. Your selected filter will now appear under the Common Metadata Filters, and the number of images that match your filter will be reduced. You can add as many filters as you want, although the Acquisition Date and the Cloud Cover are the most useful.

The next step is to view the images you have selected and choose the one you want to download using the Scene Navigator panel on the bottom left of the main image. The scene ID is at the top of this panel, and underneath is the data set. On the main screen, to the left of the latitude and longitude, is the acquisition date and time of the displayed scene.

This scene is shown as the image in color on the map. If you zoom out sufficiently, the current scene is shown with a red outline – although it is possible to change this color in the Map Layers Preferences section under User Preferences, accessed via its icon, a head and shoulders outline with a cog wheel at their shoulder.

Underneath the scene ID and data set descriptions in the Scene Navigator is a thumbnail of the selected image, together with a series of buttons. At the bottom of the panel are Previous and Next buttons to enable you to scroll through the available scenes for the chosen area. A Select Scene button allows you to add the scene to your saved scene list if you need to order them from Landsat; this is rare as most scenes are available in the archive.

Above the Previous and Next buttons are options to do the following:

- **Download** allows you to download these data to your own computer, which is described further in Section 4.9.
- **Metadata** opens up a window above the Scene Navigator panel, which gives lots of details on the attributes of the current scene in a tabular format.
- **Share** opens a new window with options to browse a reduced spatial resolution version of the whole image and the full metadata details for the scene. It also gives you the option to download the scene.
- **Hide** removes the current scene from the list of available scenes. It is helpful if you are narrowing down potential scenes but don't want to delete the scene in case you need it, so you can hide it instead. The icon looks like rubber on the Selected Data Set(s) panel and allows you to return all hidden scenes to your list of available scenes.

If you select the thumbnail image, it brings a new panel with Scene Browse options. This option lets you see small versions of the images available to download, namely the natural color (referred to as reflective browse), thermal image, and quality image. These options are described in more detail in Section 4.9 below on downloading data.

Finally, note that you have to be logged in to have all options available, and, as noted in Chapter 2, Landsat takes several days to revisit the same locations on the Earth and, therefore, the archive does not have daily images.

4.7 Scene ID

Each selected scene has a specific Landsat Scene Identifier (ID), which is visible at the top of the Scene Navigator Panel, consisting of 40 characters. For example, LC08_L1TP_203025_20210907_20210916_02_T1 is the Scene ID for an image of Plymouth, UK, in September 2021. Each element of the information in the Scene ID is separated by an underscore, such that:

- The first four characters (LC08) are the Landsat sensor and mission, where C shows it is from OLI/TIRS instruments combined, and the 08 represents Landsat-8.
- After the underscore, the following four characters are the processing correction level. In this case, Level L1TP means a high-precision geometric correction has been applied, including a terrain correction. The processing and correction options are described in more detail in Section 6.4.
- After the underscore, the following six characters are the WRS path (203) and row (025) (see Section 4.7).
- After the underscore, the following eight characters give the year (2021), month (09), and day (07) of acquisition. (This is the important date to know.)
- After the underscore, the following eight characters give the year (2021), month (09), and day (07) the data were processed.
- After the underscore, the next two digits show the collection number (01) or (02), as described in Section 4.4.
- Finally, after the underscore, the last two digits are the collection category where RT = Real-time, T1 = Tier 1, and T2 = Tier 2, as described in Section 4.4.

4.8 Worldwide Reference System

The WRS is a notation system that Landsat uses to map its orbits around the world and is defined by sequential path and row numbers. Despite its name, there are in fact two versions of the WRS: WRS-1 for Landsat missions 1–3 and WRS-2 for the rest of the Landsat missions.

The paths are a series of vertical tracks going from east to west, where Path 001 crosses the equator at 65.48° west longitude. In WRS-1, there are 251 tracks, whereas the instruments on Landsat-4 and beyond have a wider swath width and thus only require 233 tracks to cover the globe. Both WRS-1 and WRS-2 use the same 119 rows, where row 001 starts near the North Pole at latitude 80°, 1 minute, and 12 seconds North (can be written in an abbreviated form as 80°1′12″ N), row 60 coincides with the equator at 0° N, and row 119 mirrors the start at 80°1′12″ S. A combination of path and row numbers gives a unique reference for a Landsat scene; the path number always comes first, followed by the row number. For example, 203-025 is the WRS-2 path and row for Plymouth, United Kingdom.

There are maps available of the paths and rows. However, there is also a handy website from USGS (https://landsat.usgs.gov/landsat_acq#convertPathRow) that converts path and row numbers to latitude and longitude and vice versa; it's accompanied by a map so you can tell you've got the right area.

4.9 Downloading the Level 1 Landsat Data

After clicking Download, you'll be shown options for downloading the Landsat products available for the selected scene. If you are not logged in, you may be asked for your username and password at this point. This might include the following:

- Landsat Collection 2 Level-1 Product Bundle: This full-sized data set is a very large compressed file containing each spectral waveband that can be ingested in, and manipulated within, image processing software. These files will be used in Chapter 6.

- Full-Resolution Browse (Natural Color) GeoTIFF: A GeoTIFF is a specific image file containing the geographic reference information to enable the data to be uploaded into a Geographical Information System (GIS), and there is more detail on using Landsat data within a GIS in Chapter 7. The Natural Color image is the easiest to understand as the colors are similar to what your eye will see.

- Full-Resolution Browse (Thermal) GeoTIFF Image: A GeoTIFF of the thermal band, which shows the variations in temperature; the darker areas are colder, and lighter areas are warmer.

- Full-Resolution Browse (Quality) GeoTIFF Image: A GeoTIFF that shows the positions of the clouds and other features, such as snow and ice, within the scene.
- Full-Resolution Browse (Natural Color) JPEG Image: This is a simple JPEG image of the natural color image.
- Full-Resolution Browse (Thermal) JPEG Image: This is a simple JPEG image of the thermal image.
- Full-Resolution Browse (Quality) JPEG Image: This is a simple JPEG image of the quality image.

When you click on the Download icon, a downward-facing arrow on top of the computer hard disk, next to a product, will start the download of the chosen product. For this chapter's practical exercise, in Section 4.12, we will download a Full-Resolution Browse (Natural Color) Image JPEG.

4.10 Basic Viewing and Using the Landsat Data

The easiest way to view a Full-Resolution Browse Natural Color Image is to open it using the Microsoft Photos tool, or something similar, where you'll be able to see the image and zoom in and out of it. Alternatively, open the image in Windows Paint 3D, Windows Paint, or a similar drawing package, and carry out some simple image processing techniques using the basic tools provided. For example:

- Zoom into, or rotate, the image.
- Use the resize tool to make the image smaller, either by reducing the size of the image in percentage terms or by reducing the number of pixels in the image.
- Crop the image to extract a particular portion.

Landsat images are free to download, and there are no restrictions on how Landsat data can be used or redistributed. It carries no copyright; however, USGS does request that they're attributed appropriately, for example, *"Landsat-8 image courtesy of the U.S. Geological Survey"* – which means you can use Landsat images on your website or other publicity materials (https://www.usgs.gov/centers/eros/data-citation).

4.11 Landsat Known Issues

Calibration is essential for any remote sensing system and includes pre-flight and postlaunch activities because the launch process can change the calibration. For Landsat, the USGS provides the calibration files and is used within the L0 to L1 processing. However, if you download and use the Landsat images, it won't be long before you encounter known issues (also called anomalies) within the imagery. These are created during the acquisition, downlinking of the data from the satellite, or transfer between storage media and are components in the imagery that the calibration has not been able to remove.

The USGS known issues webpage (https://www.usgs.gov/landsat-missions/landsat-known-issues) gives a list of known issues, but some common ones are as follows.

4.11.1 Scan Line Corrector within Landsat-7 ETM+

Landsat-7 ETM+ suffered a failure of its Scan Line Corrector (SLC) on May 31, 2003. The SLC's role is to compensate for the forward movement of the satellite as it orbits, and the failure means that instead of mapping in straight lines, a zigzag ground track is followed. This failure causes parts of the image edge not to be mapped, giving black stripes of missing data between 390 and 450 m (i.e., 13–15 pixels) in size, which the USGS estimates causes a loss of approximately 22% of the scene's data; an example of this issue is shown in Figure 4.2. On the "Sensor/Product" section of the Scene Information, the notation SLC-off indicates that a Landsat-7 ETM+ image

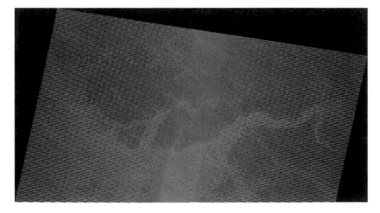

FIGURE 4.2
Landsat-7 ETM+ scene from December 10, 2003, displayed as waveband 1 shown for the whole scene with the SLC artifacts. (Data courtesy of USGS/ESA.)

was taken after the SLC failed. Further details can be found on the USGS Landsat-7 webpage (https://www.usgs.gov/landsat-missions/landsat-7) or within Storey et al. (2005).

4.11.2 Bright Pixels

Another example is brightly colored single pixels that don't match their surrounding area, caused by impulse noise that can also be visible as dark or missing pixels. The known issue has several potential causes, including technical issues during the downlink or the transcription of the data from tape to digital media. Therefore, it occurs more frequently for ground stations encountering transmission issues or the more historical data sets, such as MSS, that were initially stored on magnetic tapes.

As a word of caution, bright pixels may not always be anomalies as small fires on the ground can also show up as the same effect. However, as Landsat multispectral data have a 60–30 m spatial resolution, single pixels aren't campfires or barbecues but would be high-temperature features such as brush burning, wildfires, or gas flares. Images heavily affected by impulse noise aren't released into the USGS archive, and the bright pixels are only visible when zoomed in; hence, selecting another image from a different date will most likely cure the phenomenon unless it is a long-term fire.

4.12 Practical Exercise: Finding, Downloading, and Viewing Landsat Data

This chapter has described the basic principles of finding, downloading, and then performing simple processing on that data. We're going to end with a practical exercise to check you've understood the principles.

The practical exercise is to create an image of Uluru, also known as Ayers Rock, located in the center of Australia. The first step is to locate Uluru on GloVis, and unless you know Australia well, it will be almost impossible to find it using the map alone. The latitude and longitude coordinates of Uluru are 25.344646 South and 131.035904 East; these figures can be put into GloVis – remembering that southern latitudes are entered as negative numbers and eastern longitudes are positive numbers; then press Jump To Location. The map should now show Australia's center, which can be confirmed by zooming out.

Select a Data Set, and find a scene with a low level of cloud cover, remembering it's possible to use the Common Metadata Filter Cloud

Cover percentage to limit the suggested scenes. You can also put a data range into the Common Metadata Filter, or scroll through the available images using the Previous Scene and Next Scene buttons.

For our example, we've selected Landsat 8–9 OLI TIRS C2 L1 images, acquired between 01/01/2021 and 31/12/2021 with 0% cloud cover (put 0 in both boxes). Twelve Landsat scenes match these criteria, and you'll see a reddish-brown image of the red center of Australia with Uluru under the crosshair. You can scroll through these images and choose whichever one you want, but for this exercise, we have selected the image from May 29, 2021 with the scene ID LC08_L1TP_104078_20210529_20210608_02_T1.

On the image, if you zoom in, you'll see Uluru and other rock formations to the left of Uluru, and if you put your cursor on the map it will change to a pointing finger. You can move around the same way as you moved around the map earlier, and the latitude and longitude will change to wherever your cursor is. If you zoom out, you'll see that the Uluru is at the top of the image just to the right of center.

Click on Download in the Scene Navigator panel, and you'll get options to download the available Landsat products for this scene. In this exercise, we only want the Full-Resolution Browse (Natural Color) JPEG Image, and so click on its Download icon, the downward-facing arrow on top of the computer hard disk, and you'll get a file named "LC08_L1TP_104078_2021 0529_20210608_02_T1.jpeg" downloaded to your computer.

If you open the file in Microsoft Photos Paint, or a similar package, you should be able to see the full scene. We know that Uluru is toward the top right corner, as shown in Figure 4.3, and if you zoom into this area, Uluru should reveal itself clearly as shown within the zoomed-in area; a little to the left is the other major red center landmark, the Kata Tjuta rock formation.

Producing images, like the one of Uluru, can easily be done using any package with a zoom and crop function; we've used Microsoft Paint 3D, but any similar drawing package would be suitable.

When the image is opened in Microsoft Paint 3D, it is very large, so all you may see is the black border of the image; however, by moving around and zooming in and out, the area of interest can be found. Once you've zoomed into the smaller part of the image you want to extract, press the Select button, and select the area of interest by drawing around it, and then press the Crop button to remove the chosen area from the image. If you zoom into the area you are interested in making the image bigger, but be careful not to zoom in too much, as the image will become blurry because of the pixel size.

Now that you've found one image using GloVis and Landsat, you can repeat the exercise anywhere else in the world – or with other Data sets. In the next chapter, we'll discuss more of the theory behind remote sensing

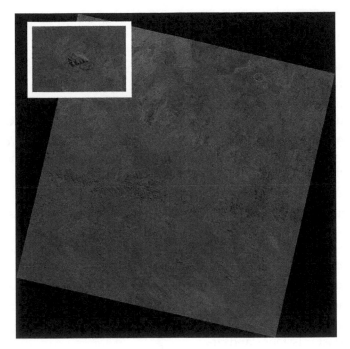

FIGURE 4.3
Landsat-8 image of central Australia acquired on March 10, 2021, with the approximate position of Uluru marked on the whole scene and a zoomed-in area to show Uluru in detail. (Data courtesy of the U.S. Geological Survey.)

and image processing regarding how the images are created and how you can manipulate them. Then, in Chapters 6 and 7, we'll give you more step-by-step instructions to perform your own image processing.

4.13 Summary

This chapter has introduced your first remotely sensed data set. It has given a summary of the Landsat missions and the type of the data they've collected. We've also begun hands-on remote sensing and have used Landsat as an illustration of how to find, download, and view satellite data.

The exercise at the end of the chapter offered an opportunity to test your understanding of the first steps of remote sensing, in preparation for future practical exercises.

4.14 Online Resources

- Associated learning resource website: https://playingwithrsdata. com/
- Converting WRS path and row numbers to/from latitude and longitude: https://landsat.usgs.gov/landsat_acq#convertPathRow
- ESA Landsat archive of MSS/TM/ETM+ products: https://earth. esa.int/eogateway/search?text=&category=Data&filter=landsat& subFilter=data%20description&sortby=RELEVANCE
- GloVis website: https://glovis.usgs.gov/
- Landsat anomalies: https://www.usgs.gov/landsat-missions/ landsat-known-issues
- Landsat Collection 2: https://www.usgs.gov/landsat-missions/ landsat-collection-2
- Landsat acquisition calendar: https://landsat.usgs.gov/landsat_acq
- Landsat copyright: https://www.usgs.gov/centers/eros/data-citation
- Landsat-7 ETM+ SLC failure: https://www.usgs.gov/ landsat-missions/landsat-7
- Landsat IGS Network: https://landsat.usgs.gov/igs-network
- Landsat mission details: https://www.usgs.gov/landsat-missions/landsat-satellite-missions/
- Useful Landsat web tools: https://www.usgs.gov/landsat-missions/landsat-tools
- USGS Landsat archive: https://landsat.gsfc.nasa.gov/data/ where-to-get-data/

4.15 Key Terms

- Scan line corrector: Failed for Landsat ETM+ on May 31, 2003 and causes parts of the swath edge not to be mapped, giving black stripes that are missing data in the affected Landsat scenes.
- Worldwide Reference System: What Landsat uses to define its orbits around the world and is defined by sequential path and row numbers.

References

Campbel, J. 2015. Landsat seen as stunning return on public investment, Blog, USGS. Available at http://www.usgs.gov/blogs/features/usgs_top_story/business-experts-see-landsat-as-stunningreturn-on-public-investment/ (accessed April 17, 2015).

Storey, J., P. Scaramuzza, G. Schmidt and J. Barsi. 2005. Landsat 7 scan line corrector off-gap filled product development. In *Pecora 16 Conference Proceedings*. Sioux Falls: American Society for Photogrammetry and Remote Sensing. Available at http://www.asprs.org/a/publications/proceedings/pecora16/Storey_J.pdf.

Wulder, M. A. and N. C. Coops. 2014. Make earth observations open access. *Nature* 513:30–31.

5

Introduction to Image Processing

This chapter focuses on the underlying theory of image processing so that you'll be able to understand the techniques used in the following chapters when they will be applied to remote sensing data within image processing and Geographic Information System (GIS) packages.

5.1 What Is an Image and How Is It Acquired?

At its simplest, an image is a series of numbers stored in columns and rows, known as a two-dimensional (2D) array, where the pixels make up the image and their values represent the shades of gray or color of the pixel as it appears on the screen. A pixel may be shown as a point (circle) or a square, and the individual pixels will subsequently become visible by zooming into an image. In image processing, data stored in this manner are referred to as raster data as they are a rectangular matrix. It can be contrasted to the way shapes are stored in GIS software, known as vector data, which are composed of points, lines, or polygons defined by their shape and geographical position. Neither of these ways of storing data actually matches the reality of the surface of the Earth that is three-dimensional, and thus, image projection methods are needed to define the real-world shape and position of a pixel. These will be discussed in detail in Chapter 7.

Satellite sensors use a number of different methodologies to capture the data within an image. Optical images are normally captured by a line scanning sensor using a pushbroom or whiskbroom technique. The pushbroom technique is where the sensor acts like a broom head and collects data along a strip as if a broom was being pushed along the ground, with the width of the image being defined by the number of pixels across the "broom head". The number of rows of data collected depends on how long the broom was pushed. In contrast, the whiskbroom approach involves several linear detectors (i.e., broom heads) perpendicular (at a right angle) to the direction of data collection. These detectors are stationary in the sensor, and a mirror underneath sweeps the pixels from left to right (see Figure 5.1) reflecting the energy from the Earth into the detectors to

DOI: 10.1201/9781003272274-5

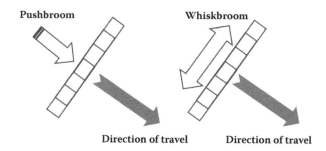

FIGURE 5.1
The difference between pushbroom and whiskbroom scanners.

collect the data. All Landsat sensors before Landsat-8 used the whisk-broom design, while OLI and OLI2 have pushbroom designs. Other push-broom systems include Satellites Pour l'Observation de la Terre (SPOT) or Sentinel-2's Multispectral Imager, while Moderate Resolution Imaging Spectroradiometer (MODIS) uses a whiskbroom system.

The microwave Synthetic Aperture Radar (SAR) system, as described in Chapter 3, uses a synthetic antenna of around 10 m in length rather than a long physical antenna. All the recorded reflections for a particular area are processed together, into an image held in a 2D array, which requires significant processing to end up in that format.

Interferometric SAR (InSAR) is a type of SAR that uses phase measurements from two or more successive satellite SAR images to determine the Earth's shape and topography, calculate Digital Elevation Models (DEMs), or measure millimeter-scale changes in the Earth that can be used to monitor natural hazards as described in Section 8.5.

The interferometry principle combines separate EM waves by super-imposing them, where two waves with the same frequency result in a pattern determined by their phase difference. Therefore, in-phase waves experience constructive interference (combined waves increase in ampli-tude) and out-of-phase waves undergo destructive interference (combined waves decrease in amplitude) (see Figure 5.2). This type of processing is needed for the Soil Moisture and Ocean Salinity (SMOS) satellite, which has 69 small receivers that measure the phase difference of the inci-dent radiation. Its Y-shaped antenna allows the spatial resolution to be improved, and the imaged area is a hexagon-like shape approximately 1,000 km across. See the European Space Agency (ESA) SMOS mission video for an introduction, which was recorded before SMOS was launched (https://www.esa.int/esatv/Videos/2009/07/SMOS_Mission).

Another acquisition method is where data are collected as individual points, such as for the traditional LRM microwave altimetry approach, so it only has one dimension (see Chapter 10 for further details).

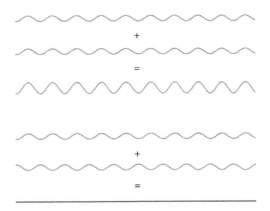

FIGURE 5.2
Constructive interference occurs when the waves are in phase, whereas destructive interference occurs when the waves are out of phase.

However, when several points are collected over a period (days for altimetry) and are brought together and gridded into rows and columns, a single 2D image is created; altimetry is described further in Section 10.2.

It's worth noting that there's also a third dimension beyond rows and columns. For optical instruments, such as Landsat, this is the number of spectral wavebands; for microwave sensors, this is the different polarizations or angles. The third dimension can also be time. When both exist, we have a four-dimensional data set that starts to become difficult to visualize, so there are now technologies called data cubes to support this (see Section 13).

5.2 Image Properties

Two terms useful for understanding the properties of image data are *radiometric resolution* and *signal-to-noise ratio* (SNR), which are linked to the sensitivity of the sensor, that is, the magnitude of the variations in emitted or reflected energy the sensor can detect.

Radiometric resolution refers to the range in brightness levels that can be applied to an individual pixel within an image, determined on a grayscale – all satellite images start off in black and white, and it's only when several layers are combined, or a color palette is added that colors appear. The maximum range of the gray values depends on how many binary numbers, known as bits, are used to store the brightness levels; the higher the number of bits used, the more detail an image has. For example, for an image recorded with a radiometric resolution of 8 bits (or 1 byte), each pixel has the potential to have 2^8 (or 256) possible different brightness levels, from 0, which indicates that the pixel will be displayed as black, up to 255, which indicates that the pixel will be displayed as white. Figure 5.3a is a single spectral waveband of Landsat Thematic Mapper (TM) used to show variations in these gray levels.

The number given to each pixel is known as its Digital Number (DN), and the histogram of the DN values for Figure 5.3a is in the top left corner of the image. A histogram is a plot of the DN values on the x-axis with the number of pixels having each value on the y-axis; the number of zero values cannot be seen because they coincide with the y-axis. The maximum DN value on this image is 245, but the histogram shows that most DN values are found between 41 and 73, although this precision may not be obvious from the figure; the precise figures are available in the image processing packages as described in Chapter 6. As most of the pixels are between 41 and 73, the image has a low dynamic range; that is, the pixels are not spread out along the full possible range of DN values. The dynamic range can be increased, by stretching the histogram, described further in Section 5.4.

The SNR is the amount of usable information compared with background noise in the signal. A low value for the SNR would be around 300:1, and a high value would be around 900:1; higher values are better because there is a greater amount of usable information in the signal. SNR is particularly important when sensing targets with low reflectance, such as water.

Landsat-8's primary mission is detecting features on land; therefore, the data over the ocean will be noisy as it has a poor SNR. In contrast, sensors designed to monitor the oceans, such as MODIS, will have a higher SNR for their ocean color wavebands.

5.3 Why Are Remotely Sensed Images Often Large in Size?

If an image has a matrix of 512 rows by 512 columns in size (in image processing, rows are used to refer to the x-direction dimension, and columns to the y-direction), then the overall image size would be $512 \times 512 = 262{,}144$ pixels.

As noted in Section 5.2, each pixel has a DN, which requires 8 bits of storage, to show the range of 0 to 255, then the total amount of storage required for that image is $262{,}144 \times 8 = 2{,}097{,}152$ bits. In computer storage terms, there are 8 bits in a byte, meaning it takes 262,144 bytes to store an image, equating to 0.25 megabytes (MB) of data storage to hold one small image.

Some satellites have a greater brightness range; for example, Landsat-8 can distinguish between over 65,000 different levels of brightness.

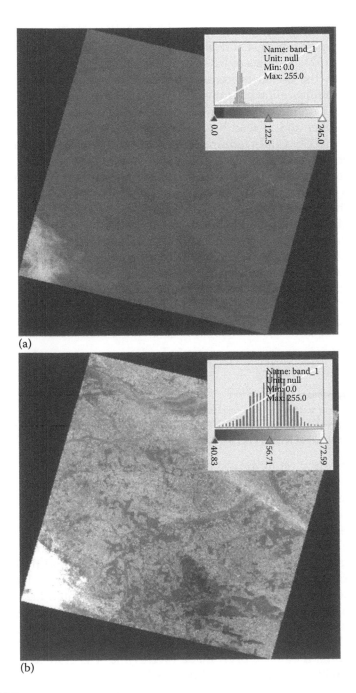

FIGURE 5.3
(a) Landsat-5 Thematic Mapper waveband 1 scene over Europe (path 188, row 024) with the accompanying image histogram and (b) contrast-stretched version with accompanying histogram. (Data courtesy of ESA/USGS.)

These images require 16 bits to store each pixel's DN. Therefore, to store a 512 × 512 pixel image, Landsat-8 requires 512 × 512 × 16 = 4,194,304 bits, equating to 524,288 bytes or 0.5 MB.

However, the actual processing itself often requires various mathematical formulas, allowing images to be added together, multiplied, subtracted, and divided. When this is done, the DN is usually no longer a whole number and instead has several decimal places such as "10.267". Storing these DN values requires at least 32 bits per pixel, known as floating point values, and gives a brightness value range of -3.4^{38} to $+3.4^{38}$. Therefore, to store a 512 × 512 pixel image now requires 512 × 512 × 32 = 8,388,608 bits, equating to 1,048,576 bytes or 1 MB.

We've been using a small image for these calculations. Let us scale this up to a single Landsat-4 TM image, which has 7,291 × 7,991 pixels using 32-bit values in each of seven spectral bands; storing this image requires 7,291 × 7,991 × 32 × 7 = 13,050,773,344 bits, equating to 1,631,346,668 bytes or 1,556 MB or 1.5 gigabytes (GB)!

In addition, Landsat-4 has fewer spectral bands than subsequent missions, for example, Landsat-8 has 11 wavebands. This demonstrates why so much disk space is needed when undertaking practical remote sensing exercises.

5.4 Image Processing Technique: Contrast Manipulation/Histogram Stretching

One of the simplest forms of image enhancement is to adjust the brightness and contrast by applying contrast manipulation, which is the same technique used when the brightness and contrast settings on a TV or computer monitor are adjusted.

As noted in Section 5.3, images often don't use the entire range of brightness available to them. Therefore, it's possible to stretch the brightness histogram by spreading the pixels out along the x-axis and thus increasing the range of brightness levels used within the image. This means some pixels become darker while others become brighter. Figure 5.3a is the original Landsat scene, and histogram, and then in Figure 5.3b, the contrast-stretched scene is shown alongside the resulting histogram. In Figure 5.3b, the cloud pixels in the bottom left have become saturated, which means they have been made bright, and it is challenging to see variation within this part of the image, whereas there is now much more visible variation in the DN values across the rest of the image, which gives a better contrast. By stretching the brightness histogram, we've sacrificed the detail within the cloud to provide a much better contrast across the whole image.

In Figure 5.3a, the histogram has a single brightness peak where most of the pixels are located. Figure 5.4a shows a Landsat-5 TM waveband 1 (blue/green) image for a region with both land and water with the histogram showing two distinct sets of pixels, one related to the water part of the image and the other to the land part; this is known as a two-peaked histogram. When there is more than one set of pixels, it's possible to histogram stretch the image so that particular features are enhanced; Figures 5.4b and c, respectively, show a zoomed-in region of the image that has been alternatively stretched to enhance the land and water pixels.

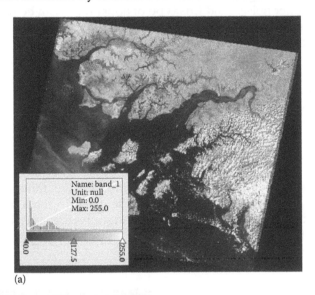

(a)

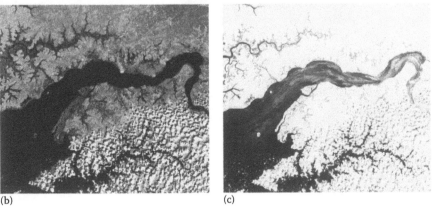

(b)   (c)

FIGURE 5.4

(a) Landsat-5 Thematic Mapper waveband 1 scene over western Africa (path 204, row 052) on December 5, 2010, having both water and land present, with the accompanying image histogram, and zoomed-in area contrast stretched to highlight the features in the (b) land or (c) water. (Data courtesy of ESA/USGS.)

5.5 Image Processing Technique: Filtering Pixels

Another image processing technique to enhance the image is filtering, where the image DN values are altered to make the image appear less noisy and smoother (more blurred) or to enhance specific features such as edges.

Images degraded by occasional pixels with very high or very low DNs have very bright or very dark pixels in parts of the image where they shouldn't be. In image processing, this is referred to as the image having "salt-and-pepper noise," and this type of appearance can be improved by applying a median filter to the image; a Landsat anomaly causing this type of effect is the "Bright Pixels" described in Section 4.9.

Median is a statistical term meaning to find the middle number in a list of numbers; hence, in the list 1, 3, 5, 7, and 9, the median is 5. Median filtering works similarly, creating a list of numbers from the central and surrounding pixels, called the kernel. The kernel is a grid of an odd number of pixels and columns, for example, 3 by 3, often written as 3 × 3, and the median of the nine numbers in the kernel replaces the center value in the output image. Figure 5.5a is waveband 1 from a Landsat-5 TM scene, received by an ESA ground station, affected by Bright Pixels. Figure 5.5b shows the visual improvement after applying a median filter. However, as the median filter changes the DNs of all the pixels in the image, it is not always appropriate to use if you're applying a mathematical processing technique afterward.

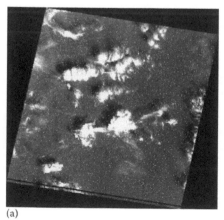 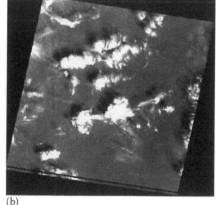

(a)   (b)

FIGURE 5.5

Landsat-5 Thematic Mapper waveband 1 scene over western Africa (path 195, row 053), affected by anomalous pixels, as what is visible (a) before and (b) after median filtering. (Data courtesy of ESA/USGS.)

Other types of filtering include the arithmetic mean filter, which averages the DNs within the kernel and uses the average DN for the central pixel; this is also known as a low-pass filter as it enhances the low-frequency spatial variations within the image. In a low-pass filter, the larger a kernel used, the greater the effect it will have on the image; in Figures 5.6a and b, respectively, the 3 × 3 and 5 × 5 arithmetic mean kernel filters have been applied and the image with the 5 × 5 kernel applied appears more blurred. The smoothing has reduced the visual influence of the salt-and-pepper noise, but not as effective as the median filter. Other applications of a low-pass filter include lowering the spatial resolution of an image, without reducing the pixel size, or subduing detailed features when image subtraction is being used to remove a background signal such as striping.

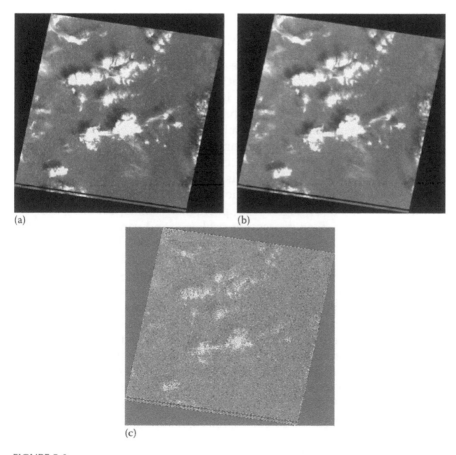

(a)   (b)

(c)

FIGURE 5.6
Mean filter with (a) a 3 × 3 kernel and (b) a 5 × 5 kernel applied to Figure 5.5a, and (c) a high-pass filter applied. (Data courtesy of ESA/USGS.)

The opposite of a low-pass filter is a high-pass filter, which enhances high-frequency spatial variations within the image and thus enhances sharp features, such as edges, within the imagery. It begins with the same averaging kernel process used in the low-pass filter. Still, instead of replacing the central pixel with the average DN, it subtracts the average DN from the central pixel's value. An example is shown in Figure 5.6c, with the filter applied to the same image as previously. The strongest spatial features in that image are the scene edges and missing lines near the bottom of the scene, which have been enhanced. In addition, the cloud edges have been enhanced slightly, and the rest of the scene is noisy.

5.6 Image Processing Technique: Applying Algorithms and Color Palettes

A key part of image processing is the application of mathematical and statistical techniques to extract information that is not obvious to the human eye, which is particularly relevant if the application involves working with multiple images or images with multiple spectral wavebands.

For example, vegetation has much higher reflectance in the near-infrared (NIR) relative to the red wavelengths, while soil and artificial materials don't. Therefore, this difference can be exploited to highlight pixels containing vegetation. Applying this to a Landsat TM image requires a mathematical calculation to be applied by dividing waveband 4 (NIR) by waveband 3 (red). Some algorithms, particularly statistical ones, are preprogrammed into image processing packages and can be applied at the touch of a button, that is, without needing to know the detail of the underlying formula. The results will often be a range of floating point numbers, for example, −203.507 to 23.167. Therefore, a histogram stretch needs to be applied to enhance what's visible.

Applying color palettes to black-and-white images is very useful as the human eye can distinguish many more colors, than shades of gray. Suppose the image has multiple wavebands/polarizations. In that case, an alternative approach is to create a color composite where three layers are displayed as red, green, and blue – the three primary colors that form the full rainbow of colors shown on a television/computer monitor. You've seen examples of these techniques in earlier chapters; if you look back to Figure 2.1a, it's a color composite, whereas Figure 2.1b has had a color palette applied.

Applying contrast stretching and color palettes requires practice, as making small changes can dramatically influence the image's appearance. The results are often subjective in terms of what one person prefers;

however, you do need to be aware of the effect this color manipulation can have on what's interpretable and how your choices affect viewers with sight problems. Some reviews are provided at https://www.gislounge.com/making-color-blind-friendly-maps/ and in Fundamentals of Data Visualization (https://clauswilke.com/dataviz/color-basics.html). These techniques will be explained further, alongside more complex algorithms, in the following chapters.

5.7 Summary

This was the second chapter focusing purely on the theory of remote sensing. We've built on some of the elements earlier in the book, and in Chapter 6, we'll be using image processing software to apply all of the techniques discussed in this chapter.

It's been more in-depth than the previous theory, but understanding the concept of Digital Numbers and how they can be manipulated within image processing is another key component to mastering remote sensing.

Although we'll have no more chapters focusing purely on theory, you aren't finished with it. All chapters in the second half will start with theory sections, before moving on to applying the relevant application techniques to data.

Your practical apprenticeship in remote sensing starts with Chapter 6; good luck, and we hope you enjoy it!

5.8 Key Terms

- Digital number: The value representing the brightness of a pixel.
- Histogram: A graph showing all the brightness levels used within an image, and the number of pixels with each brightness level.
- Kernel: An odd-numbered grid of pixels, such as 3×3 or 5×5, used to apply filters to an image.
- Radiometric resolution: The number of brightness levels within an image that defines the radiometric detail.
- Signal-to-noise ratio: The amount of useable information within a recorded signal.

6

Practical Image Processing

Chapter 5 introduced several image processing techniques used in remote sensing to enable features and information within images to become more prominent. This chapter will give a step-by-step guide to applying these basic techniques using image processing software that you can work through. It provides a foundation for the book's second half, where a number of applications will be reviewed theoretically and practically.

6.1 Image Processing Software

The manipulation of imagery, by applying the techniques described in Chapter 5, requires specific image processing software, and there are a wide range of different packages available. Various commercial packages are on the market, with license fees ranging from a few dollars to thousands of dollars. Some of the most well-known commercial packages include Adobe Photoshop, ENVI, and ERDAS Imagine. There are also several free packages available, including the GNU Image Manipulation Program (GIMP), which we've used to help create the figures in this book as it's useful for image manipulation, and ImageJ developed by Wayne Rasband at the US National Institute of Mental Health.

In addition, space agencies also provide image processing/remote sensing software to promote the use of satellite data. For example, the National Aeronautics and Space Administration (NASA) offers online web applications such as Giovanni, developed by the Goddard Earth Sciences Data and Information Services Center, and the SeaDAS desktop application, enabling users to process remotely sensed data on their own computers. Also, the European Space Agency (ESA) has several packages, including the Sentinel Application Platform (SNAP) and Bilko, which is an image processing package for education use initially developed for United Nations Educational, Scientific, and Cultural Organization (UNESCO).

The tutorials and supplementary materials in this book are based on SNAP, a stand-alone, desktop-based satellite data processing suite developed for the Sentinel satellites of the Copernicus program. The package is

DOI: 10.1201/9781003272274-6

freely available and allows the processing, analyzing, and visualizing of a wide range of remotely sensed data including Sentinel, Landsat, Moderate Resolution Imaging Spectroradiometer (MODIS), and Soil Moisture and Ocean Salinity (SMOS), among others.

Although we're using SNAP as our demonstrator, the principles and associated learning should be easily transferable to other image processing packages.

6.2 Installing the SNAP

SNAP is currently available as a series of toolboxes, and although the notation used is SNAP in the book we are referring to all the Sentinel Toolboxes, with the current (at the time of writing) version being 9.0.0. As SNAP will continue to be developed, more recent versions of the software may be available to download than the one used here. This might mean that the look, menus, and options may differ slightly. An online list of resources is available via the website accompanying this book (https://www.playingwithrsdata.com/), which will be updated to support new software versions.

The software is downloaded from the Science Toolbox Exploitation Platform (STEP, https://step.esa.int/main/download/snap-download), giving options for downloading the Sentinel Toolboxes for Windows 64-bit, Unix 64-bit (also known as Linux), or Mac OS X (Apple Mac) versions of the Toolkit.

Selecting the Main Download hyperlink for the appropriate software version for your system sets off a standard download approach. Once downloaded, the installation may run automatically, or you may need to open the downloaded file to start the installation. SNAP comes with an installation wizard; you just need to accept the default options SNAP offers during the installation.

NOTE: Like all image processing software, SNAP requires a significant amount of disk space – the compressed file for version 9.0.0 is almost 1,000 MB in size, and the final installation requires just over 1,300 MB. Therefore, it's important to ensure that there is sufficient free disk space to install the software and use it because the remote sensing data also take up significant disk space as discussed in Section 5.3. If you have difficulties using SNAP, then questions can be asked on the STEP forum (https://forum.step.esa.int/).

6.3 Introduction to the SNAP

When SNAP opens, there will be a pop-up window for SNAP Update. SNAP and the plug-ins within the software are periodically updated. The first time you open SNAP, you should select Yes and let the Update Wizard guide you through the update process. After the first time, it is helpful to check for updates regularly, which can be done through the pop-up window or Help > Check for Updates.

Once you've updated SNAP, you'll see the graphical user interface. The menu bar is running along the top of the screen; underneath is a toolbar, and on the right side are the tool windows: Product Library, Layer Manager and Mask Manager. There are three main windows – all of which will be blank when you first open SNAP:

- Image Window: This is the large main window, which displays the image you're currently working on. It can support several concurrently open images through a series of tabs, one for each image, which will appear along the top of this window.

- Products Explorer Window: This is the upper of the two panels on the left side of the screen. It contains all the information about the current images being worked on within SNAP. It has two tabs; the first is the product tab containing all the details about the products/layers within each file, while the second Pixel Info tab includes details such as the location and time of the image.

- Navigation/Color Manipulation/Uncertainty Window/World View: The lower panel, down in the bottom left corner, has four tabs.

  - "Navigation" tab shows a thumbnail version of the whole image, and zooming in and out of the main image can be undertaken using the magnifier glass icons on the navigation panel (positive for zooming in and negative for zooming out). As you zoom in, a blue box will appear on the image in the navigation panel, indicating the image's area being shown in the main window.

  - "Color Manipulation" tab allows you to manipulate the image histogram and apply colors, described in more detail in Sections 6.10 and 6.11.

  - "Uncertainty Visualisation" tab deals with uncertainty within images; it's a key focus for the ESA Sentinel missions – but we're not going to cover it in this book. If you return to the

STEP webpage you will see there are resources, such as videos, to take you beyond what we cover in this book.

- "World View" tab indicates where on the globe the image is located.

More details on how the different parts of the software work will be given as you progress through this chapter, and the book overall. The first step, before beginning practical image processing, is to download some Landsat data; Sections 6.4–6.6 will provide more details about the available data and how to get them.

6.4 The Geometry of Landsat Level-1 Data

As described in Chapter 4, Landsat data are made available at a number of levels depending on the amount of processing received. L0 is the raw data received from the satellite sensor, and it's processed by applying several algorithms to create L1 data.

The data are refined and corrected within the processing to ensure that they are as accurate as possible. The geometric processing involves confirming the position of the image and uses two external sources of data:

- Ground control points (GCPs): These are points on the image where the latitude and longitude are known, such as the corner of a building or a road.
- Digital terrain model (DTM): This is a topographical model of the planet that provides details on the undulations of the Earth's surface.

Wherever possible, the Landsat data are processed using both sources of information, and in this case, the scene is referred to as having a Standard Terrain Correction or L1TP for short. If there are insufficient GCPs available in the area, for example, snow- and ice-covered land or significant cloud cover, then the geometric correction is performed only using the DTM and information about the position of the satellite – this is referred to as Systematic Terrain Correction or L1GT for short; this option is only available for the Landsat-7 ETM+ mission. Finally, there are some scenes where neither the GCPs nor DTM are available, for example, in coastal areas or islands. Hence, the position is derived purely from what the satellite collected, referred to as Systematic Correction or L1GS for short. On the GloVis Scene Information, within the "Sensor/Product" information field, the geometric processing types are displayed for each scene along-side the sensor that collected the data.

Lastly, the data have been reprojected into the Universal Transverse Mercator (UTM) map projection. The majority of maps and atlases use a Mercator projection, to allow the spherical shape of the Earth to be printed onto a flat two-dimensional (2D) surface. UTM is a specific type of Mercator projection, which divides the world into 60 slices, each 6° wide in longitude, which ensures that the area represented by each pixel is consistent across the world.

6.5 Landsat Level-1 GeoTIFF Files

In Chapter 4, the Full-Resolution Browse (Natural Color) GeoTIFF was downloaded, and Landsat offers several other products described in detail in Section 4.7; to recap,

- Full-Resolution Browse (Thermal): An image of the thermal waveband offered both as a GeoTIFF and JPEG.
- Full-Resolution Browse (Quality): An image showing the positions of the clouds and other features, offered as both a GeoTIFF and JPEG.
- Landsat Collection 2 Level-1 Product Bundle: A large, zipped file that downloads a separate file for each individual spectral waveband measured by the sensor, plus additional information files. These files are used in this chapter and are described in more detail below.

The Full-Resolution Browse (Natural Color) Image isn't originally a single image; instead, it combines three separate spectral wavebands. Displaying the wavebands red, green, and blue produces a single-color composite image very similar to what your eye would see, and this composite is the Natural Color Image product. It's often referred to as an RGB image because of its creation from the three primary colors, red, green, and blue, that combined can generate any color – this is the approach used for televisions and monitors.

Once unzipped, the downloaded Landsat Collection 2 Level-1 Product Bundle provides a separate file for each of the individual spectral wavebands measured by the sensor and allows users to create their own images by combining the spectral wavebands in a variety of ways. The spectral wavebands are provided in the GeoTIFF format and include the geographical referencing information to allow the importing of the data into image processing or GIS software. It should be noted that the number of spectral wavebands differs depending on the Landsat mission; for example, Landsat-5 has seven spectral wavebands, whereas Landsat-8 has 11.

The full list of spectral wavebands per mission, together with the additional information, is shown in Table 6.1.

TABLE 6.1

Spectral Wavebands Offered by the Different Landsat Missions and Sensors

Landsat Missions	Instrument	Spectral Wavebands (Central Wavelength in nm)
1, 2, 3, 4, and 5	Multispectral Scanner (MSS)	• Green (550) • Red (650) • Two NIR (750 and 950)
4 and 5	Thematic Mapper (TM)	• Blue (490) • Green (560) • Red (660) • NIR (830) • Two SWIR (1670 and 2240) • TIR (11,500)
7	Enhanced Thematic Mapper Plus (ETM+)	• Blue (485) • Green (560) • Red (66) • NIR (835) • Two SWIR (1650 and 2220) • TIR (11,450) • Panchromatic
8	Operational Land Imager (OLI) and Thermal Infrared Sensor (TIRS)	• Coastal aerosol (440) • Blue (480) • Green (560) • Red (655) • NIR (865) • Two SWIR (1610 and 2200) • Panchromatic • Cirrus (1370) • Two TIR (10,895 and 12,005)
9	Operational Land Imager (OLI2) and Thermal Infrared Sensor (TIRS2)	• Coastal aerosol (440) • Blue (480) • Green (560) • Red (655) • NIR (865) • Two SWIR (1610 and 2200) • Panchromatic • Cirrus (1370) • Two TIR (10,895 and 12,005)

In addition to the spectral wavebands, the L1 data contain a varying number of additional files, which are consistent across missions for Collection 2 except for the additional Landsat-7 files:

- MTL file: Contains metadata on the data processing and calibration applied to the scene.
- GCP file: Contains the GCPs used in the geometric correction, with the residual errors for each GCP being the differences between the known location and position according to the Landsat scene.
- Geometry files: Useful when performing processing such as an atmospheric correction (see Section 2.2) where the geometry between the satellite's location, the sun, and each pixel is needed.
- Gap_mask folder for Landsat-7 files: Contains an additional image for each spectral waveband. This is attributed to the failure of Landsat-7's Scan Line Corrector on May 31, 2003; see Section 4.9 for further details. In the GloVis Scene Information Box, the "Sensor/Product" line uses the notation SLC-off to indicate that the image was taken after the SLC failed, and the gap_mask folder identifies all the pixels affected by the gaps.
- QA: The Quality Assessment files indicate the pixels that may have been impacted by external factors such as clouds. Although useful within scientific research, the file's contents can be easily misinterpreted, and thus it's not recommended for general users.
- VER.jpg (accompanied by a.txt report): Displays a grid of verification points with colors to represent the accuracy of the geometric correction.
- STAC metadata file: The SpatioTemporal Asset Catalog (STAC) family of specifications that aim to standardize the way geospatial asset metadata is structured and queried (for more information, see Chapter 13). This becomes very useful when larger data sets are stored on cloud-based storage and need to be accessed/processed using code.

Most of the additional information is either in a text file format (often called ASCII format) and, hence, can be opened and read into any basic word processing package such as WordPad or image files. Of these additional files, the MTL is the only one we'll use within the book; however, it's useful to be aware of what you've downloaded as SNAP needs some of the files when it imports the Landsat GeoTIFF wavebands; thus, they should be kept together in the same directory.

6.6 Downloading the Level-1 Product Bundle

Downloading the Collection 2 Level-1 (L1) Product Bundle data follows the same process used in Chapter 4 for the Natural Color Image; go into GloVis and select the area of the world required through the map, latitude and longitude, or use the Worldwide Reference System (WRS).

Once the selected area is correct, select the preferred Data Set, and choose the scene to download by changing the Common Metadata Filters, or using the Previous Scene and Next Scene buttons, remembering all the details about the scene are included in the Scene Information Box. Once the required image is shown, click the Download icon, which is a downward arrow on top of a computer hard disk.

The download is a compressed archive file, known as a tar file. The tar suffix indicates that it is a compressed file and will need to be unzipped using a tool such as the commercial WinZip or 7-Zip (https://www.7-zip.org/), which is freely available. Unzipping the tar file using 7-Zip gives access to the individual spectral waveband files and several other files.

For this practical, we're going to use Landsat-5 TM data covering the area around Warsaw in Poland; as these are older data, there are fewer spectral bands to deal with, so it is a bit easier to use. The first step is to move to the correct geographical location, and in this case, the chosen scene has the WRS-2 details of Path 188 and Row 24. So, enter these and select Jump To Location. Next, use the toggle buttons in the Select Data Set(s) box of the Interface Controls panel to turn on the toggle button for Landsat 4–5 TM C2 Level-1 data set, and press the Add Selected Data Set button. Finally, we will use the image acquired on October 27, 1990, so you might want to use the Metadata Filters to narrow your search down to an acquisition date during October 1990 and the Previous and Next Buttons to select the correct data. Once you have the right scene, click the Download button. From the options, click the Download icon for the Landsat Collection 2 Level-1 Product Bundle, and the large file will begin downloading. If nothing appears, it may be that your browser is blocking Pop-Ups as GloVis uses these. If you go into your web browser's settings and allow (https:\\glovis.usgs.gov\) website to create pop-ups, you will get the data.

The result of downloading the L1 GeoTIFFs should be a 175-MB tar file with the name "LT05_LT1P_188024_19901027_20200915_02_T1.tar".

You should recognize the Scene ID components from those described in Section 4.7. However, the difference between the acquired data and the processing date is worth noting. The processing date is over 25 years after it was acquired because the USGS has undertaken reprocessing activities to continually improve the data quality, which might be for a specific satellite or a subset of the entire time series or all the Landsat satellites.

The Landsat Science Team are focused on identifying and researching these improvements, and their activities are published in peer-reviewed scientific literature. Therefore, anywhere there is a significant gap between the acquired and processing dates for a scene indicates that a reprocessing exercise has taken place, and you can tell the version of the processor used from the MTL file.

Once the file has been fully unzipped, there will be the following 23 files that are a total of 175 MB in size:

- LT05_LT1P_188024_19901027_20200915_02_T1_B1.TIF
- LT05_LT1P_188024_19901027_20200915_02_T1_B2.TIF
- LT05_LT1P_188024_19901027_20200915_02_T1_B3.TIF
- LT05_LT1P_188024_19901027_20200915_02_T1_B4.TIF
- LT05_LT1P_188024_19901027_20200915_02_T1_B5.TIF
- LT05_LT1P_188024_19901027_20200915_02_T1_B6.TIF
- LT05_LT1P_188024_19901027_20200915_02_T1_B7.TIF
- LT05_LT1P_188024_19901027_20200915_02_T1_ANG.txt
- LT05_LT1P_188024_19901027_20200915_02_T1_GCP.txt
- LT05_LT1P_188024_19901027_20200915_02_T1_MTL.txt
- LT05_LT1P_188024_19901027_20200915_02_T1_MTL.json
- LT05_LT1P_188024_19901027_20200915_02_T1_MTL.xml
- LT05_LT1P_188024_19901027_20200915_02_T1_QA_PIXEL.TIF
- LT05_LT1P_188024_19901027_20200915_02_T1_QA_RADSAT.TIF
- LT05_LT1P_188024_19901027_20200915_02_T1_SAA.TIF
- LT05_LT1P_188024_19901027_20200915_02_T1_stac.json
- LT05_LT1P_188024_19901027_20200915_02_T1_SZA.TIF
- LT05_LT1P_188024_19901027_20200915_02_T1_thumb_large.JPEG
- LT05_LT1P_188024_19901027_20200915_02_T1_thumb_small.JPEG
- LT05_LT1P_188024_19901027_20200915_02_T1_VAA.TIF
- LT05_LT1P_188024_19901027_20200915_02_T1_VER.jpg
- LT05_LT1P_188024_19901027_20200915_02_T1_VER.txt
- LT05_LT1P_188024_19901027_20200915_02_T1_VZA.text

GeoTIFF files are large, both zipped and unzipped, and therefore require a lot of disk space. Despite the amount of storage space capacity on your computer, it will soon fill up if you start using remote sensing data regularly. Regular housekeeping should be undertaken to remove any files no longer needed; for example, once you've extracted the files, you can delete the tar file – you can always download it again in the future if you need it.

If you start downloading several remote sensing data sets, you'll soon have many files with similar names; good organizational file management techniques can significantly help manage this situation. We'd recommend having separate folders for each bundle of GeoTIFF files and any associated processing you do with them; we'd also suggest you rename the folder from the entity ID to help you know what it contains.

6.7 Importing Landsat Level-1 Data into SNAP

Within SNAP, it's possible to import a single-band Landsat GeoTIFF file or the whole set of available GeoTIFF spectral wavebands for a Landsat scene in one go. For this tutorial, we will import all the spectral wavebands in one go, and SNAP uses the MTL file to achieve this.

To import the data, go to the menu item File > Import > Optical Sensors > Landsat > Landsat GeoTIFF and navigate to where you have saved the data you have just downloaded, and select the MTL file called "LT05_LT1 P_188024_19901027_20200915_02_T1_MTL.txt", and press Import Product.

Once imported, the Landsat Scene ID appears at the top of the Product Explorer window. If you press the + sign on the left side of the Scene ID, the product will expand to show you what is within the scene, and the words *Metadata, Flag Codings, Vector Data, Bands*, and *Masks* will appear. They each have a small plus sign next to them, and if you press the + button next to Bands, it will expand and show seven spectral wavebands, displayed as radiance wavebands numbered one to seven. There are also flags, satflags, sun, and view data related to removing anomalous data in the image.

In the remainder of this chapter, this data set will be used to demonstrate the practical application of basic image processing techniques.

6.8 Practical Image Processing: Creating
Simple Color Composites

With the data imported into SNAP, it's possible to very quickly create some simple color composites by combining three spectral wavebands into one image. For example, combining the red, green, and blue wavebands will generate an equivalent to the Natural Color Image from Chapter 4.

To do this, click on Bands in the Product Explorer window and then go to the menu item Window > "Open RGB Image Window...", which opens a dialog box, and select "Landsat-TM 3,2,1" from the Profile drop-down menu at the top. This option will populate the RGB wavebands for Landsat-5, and clicking on OK will cause the image to appear both in the main window and in the navigation panel in the bottom left corner; it should look like what's shown in Figure 6.1a.

The scene may appear as if the pixel resolution is coarser than it should be, which is because SNAP reduces the memory requirement by using a coarser resolution to view the data initially, and then the resolution is enhanced as you zoom in.

The easiest way of zooming in and out is by using the magnifier glass icons on the navigation panel on the bottom left panel; it is positive for zooming in, negative for zooming out, P for zooming to the center of the image with the zoom factor set to the default value so that the size of an

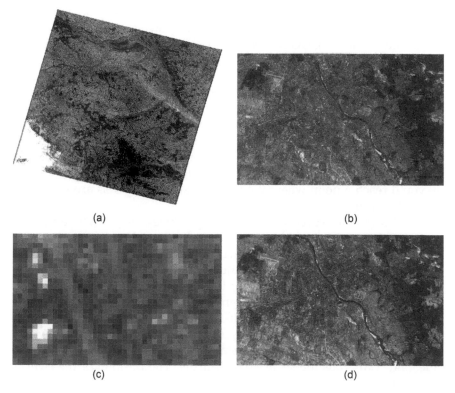

(a)　　　　　　(b)

(c)　　　　　　(d)

FIGURE 6.1
Landsat-5 Thematic Mapper scene over Europe (path 188, row 024) with the (a) pseudo-true-color composite, (b) zoomed in to show Warsaw city, (c) individual pixels when zoomed in further still, and (d) zoomed in to show Warsaw city as a false-color composite. (Data courtesy of the U.S. Geological Survey.) (NOTE: there are image tabs along the top of the main window. The original RGB image remains available when you add the false-color composite, allowing you to quickly flip between the two by clicking on the tabs.)

image pixel has the same size of a display pixel, and A takes you out to the whole scene. As you zoom in, a blue box will appear on the image in the navigation panel, indicating that the image's area is shown in the main window. You can also navigate around the image by moving this blue box.

Alternatively, on the main screen, if you put your cursor in the top left corner of the image a compass-style symbol of a white circle with arrows inside appears – called the navigation control. Clicking on the arrows will move the map in the relevant direction, and the line underneath the circle will allow you to zoom in and out by clicking left of center to zoom in, and right of center to zoom out. If you click on compass points and move your mouse, the image will rotate, with the yellow point indicating the direction of the image. Finally, clicking and holding the star in the center will allow you to move around the image quickly as you move your mouse. This navigation may take a little getting used to!

Zooming into the central top of the image shows the city of Warsaw, as seen in Figure 6.1b. This RGB composite is also known as a pseudo-true-color composite, as the image has the colors your eyes expect to see; for example, trees are green, water is blue, buildings are gray, soil is brown, and so on. However, the wavebands in optical sensors are narrower than the wavelengths picked up by the receptors in our eyes, so there will be differences.

In Figure 6.1b, the Vistula River that Warsaw straddles can be seen running through the center of the image, and the Warsaw Chopin Airport, which carries more than 10 million passengers a year, is the white area in just off-center toward the bottom of the image, and you may just be able to make out the airport runways. Similar to Figure 2.5, if you keep zooming in, you'll reach a point where the individual square pixels are visible, as shown in Figure 6.1c.

Combining other spectral wavebands produces images where the colors differ from what would be expected; these are known as false-color composites. As they use different parts of the EM spectrum, the surface of the Earth interacts differently with the radiation allowing features hidden when showing pseudo-true-color to become far more prominent. If you go to the menu item Window > "Open RGB Image Window", but this time from the dialog box select "Landsat-TM 4,3,2" in the profile drop-down menu, you'll produce a false-color composite. In this image, shown in Figure 6.1d, red indicates green vegetation, urban conurbations are green-blue, and coniferous forests are dark purple. The position of the river is more apparent, as are the airport's runways.

Using both "Landsat-TM 3,2,1" and "Landsat-TM 4,3,2" provides the two standard options for color composites; however, by changing the red, green, and blue selections in the dialog box it's possible to assign spectral wavebands of your choosing. This produces different responses from surfaces, allowing other features to be seen; the challenge is knowing what you are looking at. The use of color composites will be discussed further in the book's second half, as it's a beneficial technique and often created when trying to understand a remotely sensed data set.

It's also possible to export an image from SNAP. The menu item File > Export > Other > View as Image provides a standard file-saving dialog box. On the right side of the box are some radio buttons, with the top pair allowing you to select whether you want to save the image you can see in the main window (View Region) or the entire image (Full Scene). Underneath are Image Resolution options that enable you to use the current resolution (View Resolution), have the maximum resolution for that image that is beneficial if you want to see all the details (Full Resolution), or you can set the resolution to what you want (User Resolution). You can change the file name and save the image in various formats recognized by many other packages, allowing the images to be incorporated into reports, presentations, and web pages.

6.9 Practical Image Processing: Creating a Subset

As described in Section 5.3, files in image processing are very large. This means that it takes a lot of data storage to hold these files, both in memory and on the hard disk, and a lot of computing power to process, particularly if you're processing on your local computer and haven't got a high-specification machine. In Section 6.8, the whole Landsat scene was worked on, and the display was zoomed into the relevant part. While this works, every processing action you undertake is applied to the entire scene.

An alternative approach is to create a subset, which is a smaller subsection of the whole image. In this case, the processing is undertaken on a smaller image and doesn't require as much computing power or data storage. If your computer struggles with the processing requirements, this method should help.

To create a subset, load the whole scene and zoom into the area you want to work with; then Click on Bands and then go to the menu item Raster > Subset, and the dialog box opens. The first tab is Spatial Subset, the geographic one, which will be populated with the details for the area you have zoomed into. You'll see a view of the whole image with a blue box on the top left of the dialog box, and this will be the same as the blue box on the Navigation panel. Clicking OK will create a subset of all bands for the view area, which is the area within the light blue box on the image. A second product will now appear on the "Products Explorer" panel, which is the subset, and you can now worth with just this smaller area of the whole scene. Alternatively, it's also possible to create a subset on single wavebands or use the image metadata from the MTL file, rather or as well as the geographical area.

6.10 Practical Image Processing: Contrast Enhancement through Histogram Stretching

Using the same GeoTIFF files for Poland, as used in Section 6.7, we're only going to load in just the radiance_1 waveband at 490 nm instead of creating an RGB image. Double left-click on the "radiance_1" waveband in the layer panel; it will appear in the main window as a black-and-white image and the navigation panel in the bottom left corner.

At the bottom of the navigation panel, there are four tabs with navigation currently selected. Click on the "Color Manipulation" tab, and you'll be shown the Basic Editor. Although this is the "Color Manipulation" tab, with the image only being in black and white, this section is focused on manipulating the grayscale brightness. At the bottom, there is Range that shows the contrast stretch applied to the image, with SNAP selecting the default values it feels are most appropriate for the minimum and maximum brightness values on the image, in this case, 30.1515505–53.684601. If you change the Editor to Sliders by selecting the radio button next to it, the histogram for this waveband will be displayed. You'll again see the minimum and maximum values selected, together with the spread of the brightness values of the pixels.

On the right side of the histogram are a series of buttons. The third button down changes the histogram stretch to 2 sigma, meaning two standard deviations, which stretches 95.45% of the pixels across the full range of available brightness values. Pressing this button changes the minimum and maximum values from 29.296079 to 86.609106. The difference between the original image and the one with the 2 sigma stretch is shown in Figure 6.2a and b, respectively. The 3 sigma option stretches all the pixels across 99.73% of all brightness values (three standard deviations). Finally, the 100% option will stretch all pixels across the entire range of available brightness levels.

Using the sigma stretches or 100% buttons still only gives you limited control. However, by selecting the slider triangles on the histogram x-axis, you can move the pointers yourself and set the contrast stretch to show whatever you want. For example, suppose you move the maximum value toward the bottom of the main peak. In that case, the brightness values are narrowed to match the main histogram peak, giving a view very similar to the original stretch provided by SNAP.

However, if they're narrowed even further, the area of cloud, down in the bottom right corner, becomes saturated, and the contrast for the dark forest vegetation is improved; for example, the minimum and maximum values from 32.836028 to 36.736420 as shown in Figure 6.2c. The three magnify buttons on the right side of the histogram allow you to zoom

(a)　　　　　　　　　　　　　　　　(b)

(c)

FIGURE 6.2
Landsat-5 Thematic Mapper waveband 1 (a) as the original image, (b) with 2 sigma contrast stretch, and (c) with a customized contrast stretch. (Data courtesy of the U.S. Geological Survey.)

in and out of the histogram, horizontally and vertically, to allow more finesse with the stretching. The top button on the right side is a reset button, which takes the view back to the default histogram stretch applied by SNAP.

Using a histogram stretch needs to be experimented with, and changing the contrast can rapidly change the image. It takes practice to get a view to show you exactly what you're looking for.

An example of what stretching a histogram can show, which isn't visible without the stretch, is shown in Figure 5.4.

6.11 Practical Image Processing: Color Palettes

In Section 6.10, manipulating the grayscale brightness was worked on for single wavebands, which can improve what's visible in an image. While a computer can distinguish between hundreds of different levels of gray, the human eye is not so efficient. Images can be enhanced by adding color to them; this helps bring out features that may not have been apparent to the human eye in the grayscale version.

Again, using the same GeoTIFF files for Poland, we will use the single radiance_3 waveband (660 nm) this time. Double left-click on the radiance_3 waveband in the layer panel, and it will appear in the main window with the view shown in Figure 6.3a. To add a color palette, go

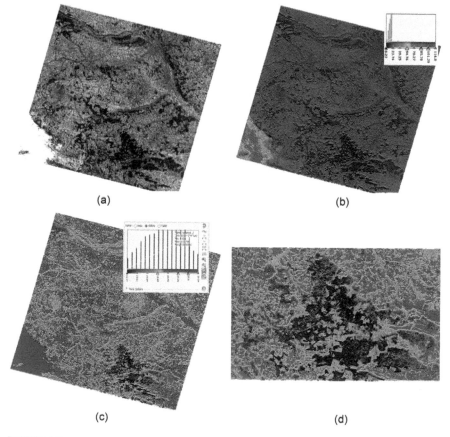

(a)   (b)

(c)   (d)

FIGURE 6.3

Landsat-5 Thematic Mapper waveband 3 (a) as the original image, (b) with the color palette applied alongside the histogram, (c) with a customized contrast stretch shown alongside the histogram, and (d) zoomed in on the coniferous forest in the bottom left corner. (Data courtesy of the U.S. Geological Survey.)

to the Color Manipulation tab in the bottom left corner window. Using the drop-down menu, the v at the end of the box, you will get a series of color palette options for the Palette. Scroll down, and you'll find the one called Spectrum, and then Click on that option. The Spectrum color palette is now applied to the image, giving a mainly blue and green view, with only the cloud bank in the bottom right corner providing any variation, as shown in Figure 6.3b. The view is mainly blue because, if you look at the histogram under the Sliders option, most of the pixels have Digital Numbers (DNs) in the blue-green range of 11.17–23.89. Applying a 95% stretch makes the view bluer, removing most of the green color. Therefore, we need to use a custom histogram stretch to bring out features within the image. Move the histogram pointers so that the selected region approximately matches the main peak of the pixels, remembering to use the histogram zoom feature to see the peak more easily. If you move the pointer on the furthest right, SNAP will automatically move the other pointers to maintain a consistent stretch. The result will be an image similar to the one in Figure 6.3c; in this image, the histogram stretch ranges from 11.17 to 28.24.

It's now possible to see a much greater color variation, with the highest DN values shown in red; mainly the clouds. As you go down through orange, yellow, and green, different types of vegetation and urban areas are visible, and if you go down to the blues, purples, and blacks, coniferous forests, and bodies of water are shown. Use the navigation panel tab to zoom in, and more detail becomes visible. For example, Figure 6.3d is zoomed in on the coniferous forest in the bottom left corner, where the clearings in the forest are visible. Still, separating cultivated fields from urban areas is difficult as both are shown with the turquoise/green/yellow/red colors.

Like histogram stretching, color palette application is an art, and it will take a bit of trial and error to get exactly what you want. You can apply these techniques to any of the spectral wavebands, and they're also used when applying algorithms, as described in Section 6.13.

6.12 Practical Image Processing: Applying a Filter

The application of filters to an image can make it appear less noisy, as described in Section 5.5, and applying filters in SNAP is relatively straightforward. As filters can only be applied to individual wavebands, we'll use the single radiance_2 waveband (560 nm); double left-click on the radiance_2 waveband in the layer panel, and the image in Figure 6.4a should appear in the main window.

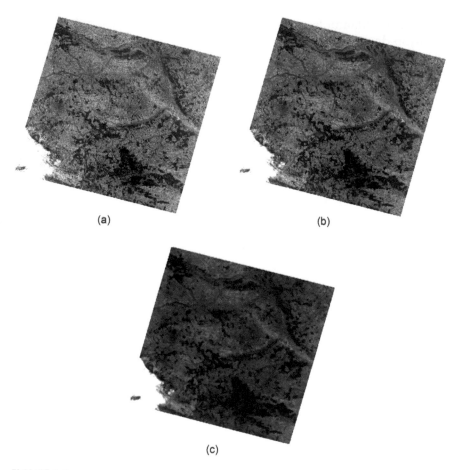

(a)   (b)

(c)

FIGURE 6.4

Landsat-5 TM waveband 2 as (a) the original image, and with a (b) 3 × 3 and (c) 7 × 7 median filter applied. (Data courtesy of the U.S. Geological Survey.)

Select the menu item Raster > Filtered Band, which shows various filtering options. A filter is selected by scrolling through the list of filters and then clicking on the one you want to choose to select it and pressing the OK button. When you select a filter, a new waveband is created in the layer panel with the filter applied. For example, in Figure 6.4b, a Non-Linear Median 3 × 3 filter is applied, whereas Figure 6.4c has a Non-Linear Median 7 × 7 filter applied. The 7 × 7 median filter has removed much more of the pixel variation, and if you zoom in, the image doesn't have the same detail in terms of the variation of the pixel DN values. Again, experimenting with the different filters, you'll see how they all work.

6.13 Practical Image Processing: Applying the NDVI Algorithm

As discussed in Section 5.6, some image processing involves the application of mathematical formulas to satellite data. As an example, this section will create a Normalized Difference Vegetation Index (NDVI) image that shows how much live green vegetation there is in an area. As vegetation has much higher reflectance in the near-infrared (NIR) than soil and artificial objects, it's possible to highlight vegetation by dividing the NIR waveband by the red waveband.

SNAP has a built-in function for calculating NDVI, making processing much more straightforward. Select the menu item Optical > Thematic Land Processing > Vegetation Radiometric Indices > NDVI Processor. A dialog box will appear with "I/O Parameters". The source product automatically selected will be the Landsat-5 data set, and the target product will be the filename and directory where the NDVI data will be saved; change the filename and output directory if it's not what and where you want. Then, move to the second tab for Processing Parameters, as you need to use the drop-down menus to select the NIR and red wavebands used to perform the calculation. Choose radiance_3 for the red source waveband and radiance_4 for the NIR source waveband, leave the Red and NIR Factors as the default values of 1.0, and then click Run. SNAP will undertake the NDVI processing, which may take a short time if you run the calculation on the whole image, and you'll be told when it is complete.

There will now be a second product in your Product Explorer window with the title LT05_LT1P_188024_19901027_20200915_02_T1_ndvi, and if the Bands are expanded for this new product, you'll see the NDVI image, which can be viewed by double-clicking on it. There will also be drop downs for flags, satflags, and NDVI flags. These are useful when performing automated processing as they can be used to filter the data. For example, the NDVI Flags indicate that if an arithmetic error has occurred, such as the output is less than zero or greater than one, then the data are considered invalid and are not used for onward processing.

The image is in black and white, and if you zoom into the river valley running diagonally across the center of the image, you'll get a view similar to the one shown in Figure 6.5a. The bright pixels indicate the presence of vegetation, and it's possible to start to see the variation in the vegetation in this area.

You can get more details by using either or both of the techniques looked at earlier within the "Color Manipulation" tab:

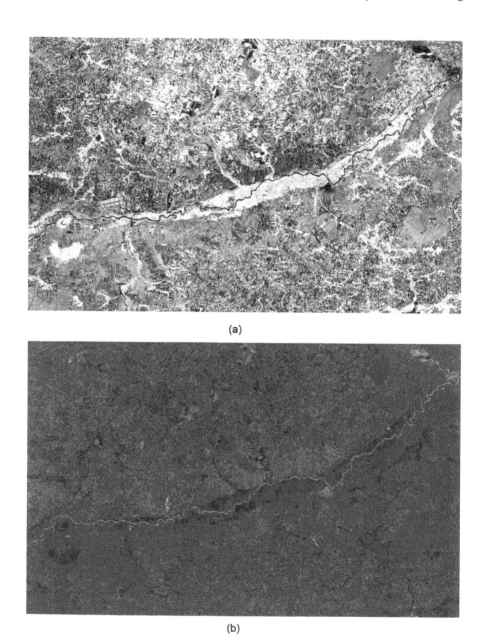

(a)

(b)

FIGURE 6.5
Calculated Normalized Difference Vegetation Index image (a) zoomed into the river valley running diagonally across the center of the image with (b) the color palette applied. (Data courtesy of the U.S. Geological Survey.)

- Color Palette: The variations can be easier to see if a color palette is added. On the Color Manipulation tab, use the Basic Editor and the drop-down menu on the Palette to scroll down to find the "meris_veg_index" option; select this by clicking on it. This option will give your view a green saturation; go to the Sliders Editor option and select the 2 sigma button to perform a histogram stretch, as shown in Figure 6.5b. The greener the area, the more vegetation is present. Interestingly, the river flood-plain areas are dark green; from looking at higher-resolution optical data in Google Earth, it's possible to see that this is a fertile area.

- Histogram Stretch: On the histogram, the positive pixels indicate vegetation; hence, you can easily see a snapshot of the vegetation to the non-vegetation. Manipulating the histogram using the outside pointers to show just the positive values will show a greater dynamic range for the vegetation.

NDVI is used in many applications, and you'll see it being mentioned in the book's second half with the theory of vegetation indices described in Section 9.2.

6.14 History of the Copernicus Program

As discussed in Chapter 4, Landsat has one of the most extended Earth Observation (EO) data archives and is one of the most popular sources of free-to-access satellite data. Another source of free-to-access data is the European Union's Copernicus Program, which operates a full, free, open data policy for all of its Sentinel satellites and derived products.

The origins of the Copernicus Program began in 1998 when the European Union signed the Baveno Manifesto, which proposed creating a European environmental monitoring program; initially called Global Monitoring for Environmental Security (GMES). In 2004, the European Commission signed a contract with ESA to establish a satellite EO component of GMES. The program was renamed Copernicus in 2012, after the Polish astronomer Nicolaus Copernicus. The following year, the EU announced that all Copernicus data would be available free-to-access to everyone.

Unlike Landsat, which operates with a single satellite for each mission, Copernicus operates multiple satellites with two initially launched for each mission that orbits the Earth 108 degrees apart to increase temporal resolution. After a few years, an additional two satellites are added. The Copernicus data collected covers the ocean and land, unlike Landsat, which focuses on the land. The first of the Sentinel satellites was Sentinel-1A, launched on April 03, 2014. To date (2022), seven Sentinel satellites have been launched, and more are planned in the coming years.

The program has been successful despite a few issues for Sentinel-1. In 2016, one of the solar wings of Sentinel-1A was struck by a millimeter-sized particle causing slight changes to the orientation and orbit that haven't impacted performance. In comparison, Sentinel-1B suffered an anomaly on December 23, 2021, causing its SAR instrument to be turned off. Unfortunately, despite their best efforts, ESA officially announced in August 2022 that Sentinel-1B's mission had ended as it was impossible to resolve the problem.

The data for Copernicus can be downloaded from the Copernicus Open Access Hub. Still, the data are also available via GloVis and other locations such as the Copernicus Data and Information Services (DIAS) that provide cloud-computing access alongside data, as do commercial cloud-computing providers such as Amazon Web Services, Google Earth Engine, and the Microsoft Planetary Computer.

6.14.1 Summary of Sentinel Missions

6.14.1.1 Sentinel-1A and 1B

The twin Sentinel-1 satellites are radar satellites carrying a 12 m long C-band Synthetic Aperture Radar (SAR) instrument, which has several modes. The most used of the modes is the Interferometric Wide swath mode with a spatial resolution of 5 × 20 m and a swath width of 250 km. Sentinel-1A was launched on April 03, 2014, and Sentinel-1B on April 25, 2016. Together they can image the entire Earth every 6 days, with a faster revisit time closer to the Equator. However, as previously mentioned, 1B stopped generating data in December 2021. So, to resume this revisit frequency, we are awaiting the expected Sentinel-1C launched in 2023.

6.14.1.2 Sentinel-2A and 2B

Sentinel-2 are optical satellites, and they both carry an identical wide-swath high-resolution Multispectral Imager (MSI) instrument with 13 spectral bands:

- Four visible and NIR spectral bands with a spatial resolution of 10 m.
- Six shortwave infrared spectral bands with a spatial resolution of 20 m.
- Three atmospheric correction bands with a spatial resolution of 60 m.

Sentinel-2A was launched on June 23, 2015, and Sentinel-2B was launched almost 2 years later on March 07, 2017.

6.14.1.3 Sentinel-3A and 3B

The Sentinel-3 satellites mainly focus on ocean measurements, including measuring sea surface height, sea surface temperature, ocean color, surface wind speed, sea ice thickness, and ice sheets. They also collect data over land, providing vegetation indices like NDVI, together with measuring the height of rivers and lakes and helping monitor wildfires.

Each satellite carries four instruments:

- Sea and Land Surface Temperature Radiometer (SLSTR) has nine spectral bands with a spatial resolution of 500 m for visible/NIR wavelengths and 1 km for thermal wavelengths.
- Ocean and Land Colour Instrument (OLCI) has 21 spectral bands (400–1,020 nm) focused on ocean color and vegetation measurements. All bands have a spatial resolution of 300 m with a swath width of 1,270 km.
- Synthetic Aperture Radar Altimeter (SRAL) that has dual frequency Ku- and C- bands. It offers 300 m spatial resolution after processing.
- Microwave Radiometer (MWR) dual frequency at 23.8 and 36.5 GHz, which is used to derive atmospheric column water vapor measurements for correcting the SRAL instrument.

Sentinel-3A was launched on February 16, 2016, and Sentinel-3B was placed into orbit on April 25, 2018.

6.14.1.4 Sentinel-5P

Sentinel-5P, or more accurately Sentinel-5 Precursor, was launched on October 13, 2017, and is dedicated to monitoring our atmosphere. It creates maps of the various trace gases, aerosols, and pollutants in our atmosphere. It carries the Tropospheric Monitoring Instrument (TROPOMI), which is a pushbroom imaging spectrometer covering a spectral range from ultraviolet and visible (270–495 nm), NIR (675–775 nm), and shortwave infrared (2,305–2,385 nm). The spatial resolution is 7 km × 3.5 km, and it has a wide swath width of 2,600 km, meaning it can map almost the entire planet daily.

6.14.1.5 Sentinel-6

Sentinel-6's full name is Sentinel-6 Michael Freilich, and it was launched on November 21, 2020. It is an oceanography satellite developed between

ESA, NASA, NOAA, EUMETSAT, and CNES (French Space Agency). The satellite carries two main instruments:

- POSEIDON-4 is a pulse altimeter in the Ku- and C- bands measuring sea level by sending microwave pulses to the ocean surface and recording how long they take to return.
- AMR-C is a multi-channel radiometer for high-resolution retrieval of water vapor content over the global and coastal oceans.

The satellite can map 95% of ice-free sea surfaces every 10 days.

6.15 Practical Exercise: Finding, Downloading, Processing, and Visualizing Sentinel-2 Data

Sentinel-2 data are similar to Landsat data, but the spatial resolution is higher. To give you an initial experience of Sentinel-2 data, we'll do a practical exercise that repeats most of the image processing you've just done with Landsat. This approach will also help reinforce learning and skills.

6.15.1 Downloading the Sentinel-2 Data

We're again going to use GloVis to download the Sentinel-2 data, and although the process is similar to what you did in Section 6.6, there are some differences.

We will use Bangkok, the capital of Thailand, in this exercise. Use the "Jump To – Click for options to jump to a specific location/scene" icon that looks like a map and compass to get to the latitude and longitude options, and then enter 13.7525 North and 100.49417 East, which means both are entered as positive numbers. Then, select Jump To Location. Once you are over Bangkok, go to the Selected Data Set(s), turn on the toggle button for Sentinel-2, and click Add Selected Data Sets. Use the Common Metadata filters to set the Cloud Cover to 0% to 20%, the acquisition date range from January 01, 2020 to March 31, 2020, and press Close. This filtering should bring up 21 scenes that match the criteria.

Scroll through these images, and you'll see that some have white clouds obscuring what is beneath. You'll also notice a jagged black line on some images, indicating a partial image available; this is due to the varying orbit of the Sentinel-2 satellite, meaning that sometimes the captured swath is not positioned over an area to acquire a full tile image. So only part of the selected area is acquired. On GloVis, it often puts these partial images on top of other full tile images, so ensure that you pick an image that does not have a jagged black line on it, which can be seen on the thumbnail image.

For this exercise, we're going to use the image for January 27, 2020 – although you should see two images from that date, one is a partial image and so is primarily black, as you'll see on the thumbnail. We don't want to use that one, so choose the full image. Scroll through to this image and then click the Download icon of a black arrow on top of a computer hard disk. You'll have two options, different from the Landsat options previously, to download the "LC1 Tile in JPEG2000 format' or the 'Full-Resolution Browse in GeoTIFF format". The first of these options is similar to the Landsat Product Bundle we used earlier in the chapter, while the second is equivalent to the Landsat Natural Color version we used in Chapter 4.

Download the "LC1 Tile in JPEG2000 format" option by clicking the button. There can sometimes be a short delay as you connect to the server to download the image; don't worry, waiting is something you'll need to get used to working with satellite data.

The result of downloading the "LC1 Tile in JPEG2000 format" will be an 820-MB zip file with the name "L1C_T47PPR_A015012_20200127T035405.zip". This filename is also the Scene ID from GloVis, which is different from the Landsat one.

- The first three characters (L1C) indicate the data level; in this case, Level 1C is a Level 1 product with radiometric and geometric corrections, including orthorectification and spatial registration.
- After the underscore, the following five characters (T47PPR) are the Sentinel-2 tile number, which is a specific identifier for each tile. The tiles are 100 km by 100 km squared in size, and neighboring tiles overlap by 10 km.
- After the underscore, the following seven characters (A015012) are an A to indicate that it is the absolute orbit number, and the six numbers indicate the orbit itself.
- After the underscore, the remaining 15 characters show the following:
  - the year, month, and day of acquisition for the first eight characters (20200127).
  - then after the T, the time in hours, minutes, and seconds (T035405) in Coordinated Universal Time (UTC) that all time zones are referenced with respect to.

As this is a zip file, it will need to be unzipped, creating a folder called "S2B_MSIL1C_20200127T034019_N0208_T47PPR_20200127T071054.

SAFE" which will contain over 100 files. This folder is a SAFE (Standard Archive Format for Europe) format ESA generate the files.

The name of this folder is the Sentinel-2 scene ID from ESA. This naming can be slightly confusing as USGS and ESA use different identifiers, but the information is similar, such that:

- The first three characters (S2B) indicate the satellite that acquired the data; in this case, it was Sentinel-2B.
- After the underscore, the following six characters (MSIL1C) show the instrument that acquired the data, which was the MSI, and the processing level of the data, that is, Level 1C (L1C).
- After the underscore, the following fifteen characters (20200127T034019) show the following:
  - the year, month, and day of acquisition for the first eight characters (20200127).
  - then after T, the time in hours, minutes, and seconds (T034019) in Coordinated Universal Time (UTC).
- After the underscore, the following five characters (N0208) show the processing baseline number to identify the version of the processor used.
- After the underscore, the following five characters (T47PPR) are the Sentinel-2 tile number.
- After the underscore, the remaining fifteen characters (20200127T071054):
  - are the year, month, and day of processing (20200127).
  - after T, the time in hours, minutes, and seconds (T071054) in UTC.

In the unzipped folder, there will be five further folders and four files:

- AUX_DATA folder containing the auxiliary data used in the processing, such as the ozone and water vapor concentration derived from a meteorological model.
- DATASTRIP folder containing details on the chunks of data as downloaded by one, or more receiving stations, and used to generate the tile. This folder includes quality assurance data in the QI_ DATA subfolder.
- GRANULE folder containing first a tile-specific subfolder. This folder has the image files, in JPEG 2000 that is a compressed format, in the IMG_DATA subfolder, with tile-specific auxiliary and quality assurance in the AUX_DATA and QI_DATA folders.

- HTML folder containing information about the product in an HMTL and associated files for online viewing.

- rep_info folder containing the format definitions for the different file types in XML format.

- 00Readme_Sentinel_Data_Terms_and_Conditions file containing a legal notice on what you can do with Sentinel data.

- INSPIRE XML file containing metadata conforming to the INSPIRE (Infrastructure for Spatial Information in the European Community) Metadata regulation, useful for online cataloging.

- Manifest.safe XML file containing a description of the SAFE contents, useful for online cataloging.

- MTD_MSIL1C XML file containing lots of useful metadata that can be read like a text file, and this is the important one for loading the data into SNAP!

6.15.2 Importing Sentinel-2 Level-1 Data into SNAP

As SNAP was specifically developed to work with Sentinel data, it's easier to import these data than the Landsat files. Simply go to menu item File > Open Product and select the MTD_MSIL1C file from the SAFE directory you've just unzipped, then press the Return button on the keyboard, and SNAP will import the file.

Once imported, the Sentinel Scene ID appears at the top of the Product Explorer window. If you press the + sign on the left side of the Scene ID, the product will expand to show you what is within the scene: *Metadata, Vector Data, Tie-Point Grids, Bands,* and *Masks* will appear. They each have a small plus sign next to them, and as before, if you press the + button next to Bands, it will expand and show thirteen spectral wavebands (as Sentinel-2 has more spectral wavebands than Landsat-5), plus sun, view, and masks, which relate to the removal of any anomalous data in the image. SNAP displays these as radiance wavebands numbered one to twelve – but note that there is a Band 8 and a Band 8A for the NIR bands, which makes up the 13 spectral bands of the satellite.

6.15.3 Practical Image Processing: Creating Simple Color Composites

As before, to quickly create a simple color composite, click on Bands in the Product Explorer window and then go to the menu item Window > "Open RGB Image Window…". This option opens a dialog box with "Sentinel-2 MSI Natural Colors" already selected, with the RBG wavelengths of B4, B3, and B2 pre-populated. Click on OK, and the image view will appear in the main window, as shown in Figure 6.6a.

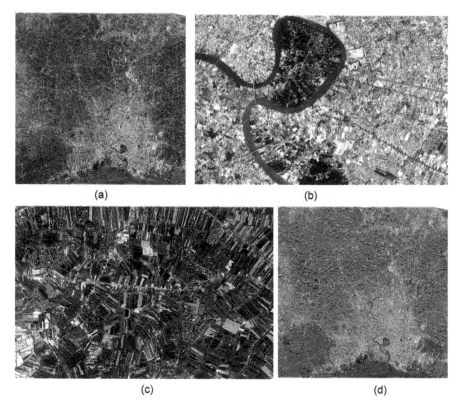

(a)   (b)

(c)   (d)

FIGURE 6.6
Sentinel-2 scene over Bangkok (tile T47PPR) with the (a) pseudo-true-color composite, (b) zoomed in over the city, (c) zoomed in on fields to the right of the city, and (d) false-color composite. (Data courtesy of Copernicus/ESA.)

The densely populated city of Bangkok is shown centrally at the bottom of the view, with the Chao Phraya River winding through the center. Equally interesting, to the city's right, are lots of rectangular farming fields with varying colors from brown and green to purple. These can be seen in Figure 6.6b and c, respectively.

A false-color-composite can be created by going to the menu item Window > "Open RGB Image Window...". This option opens a dialog box, click on the v at the end of "Sentinel-2 MSI Natural Colors" to dropdown other options. Select "Sentinel-2 MSI False-color Infrared", which will select B8, B4, and B3 for the wavebands. Click on OK, and the false-color view will appear, as shown in Figure 6.6d. The river is more evident in this view, and there is a lot of red showing the agricultural activities. The solid red areas are actively growing crops, such as rice or forest, with a bright pink area within the bend of the river being a well-water golf course also having lush green vegetation.

6.15.4 Practical Image Processing: Applying the NDVI Algorithm

With agriculture and farming on the right side of the view, it will be interesting to apply the NDVI algorithm we used earlier. We'll use all the techniques you've already learned in this chapter to apply the NDVI:

- Create a subset: Zoom into the fields on the right side of the image, similar to Figure 6.6c, then Click on Bands and then go to the menu item Raster > "Subset". Use the Spatial Subset for the selected geographic area and click OK. The subset will appear as a second product on the "Products Explorer" panel.

- Apply the NDVI algorithm: Click on the + symbol next to the new subset product in the Product Explorer Panel – use the minus (–) symbol to close the original imported data if you can't see the new product. Select the subset Bands, then select the menu item Optical > Thematic Land Processing > Vegetation Radiometric Indices > NDVI Processor. A dialog box will appear with "I/O Parameters". Change the filename or output directory if you want, and then move to the second tab for Processing Parameters. You will need to use the drop-down menus to select the NIR and red wavebands that will be used to perform the calculation. Choose B4 for the red source waveband and B8 for the NIR source waveband, and then click Run. SNAP will undertake the NDVI processing. There will now be a third product in your Product Explorer panel window with _ndvi at the end. Select this product, and use the + to show the contents. Press the + next to Bands to see the NDVI, which can be viewed by double-clicking on it. This now gives the black-and-white view like that shown in Figure 6.7a.

- Color *Palette*: Go to the Colour Manipulation tab with the Basic Editor selected, and from the drop-down list on Palette, choose the "meris_veg_index.cpd" option. This option will give your image a green/yellow saturation similar to Figure 6.7b, with darker green color indicating vegetation growth.

6.16 Summary

You've done some proper image processing in this chapter. By applying the techniques described in Chapter 5 in image processing software, you've been loading and viewing data, manipulating image colors through histogram stretches and color palettes, applying filters, and applying the NDVI algorithm.

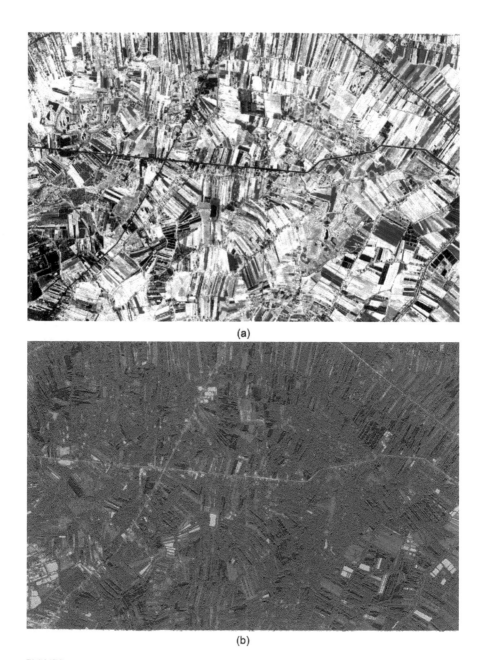

(a)

(b)

FIGURE 6.7

Calculated Normalized Difference Vegetation Index image (a) zoomed into fields with (b) the color palette applied. (Data courtesy of the Copernicus/ESA.)

These skills are the basic building blocks of remote sensing, and you'll be using them regularly throughout the rest of the book. Having learned these techniques, the best way to develop your knowledge and understanding is to plunge straight in and start to experiment with processing Landsat and Sentinel data. Go on, try it out!

The second half of the book will detail some of these techniques and highlight how they're used in real-world applications.

6.17 Online Resources

- 7-Zip: https://www.7-zip.org/
- Associated learning resource website: https://playingwithrsdata. com/
- Bilko: https://eo4society.esa.int/resources/bilko/
- GIMP: https://www.gimp.org/
- GloVis: https:\\glovis.usgs.gov\
- Science Toolbox Exploitation Platform: https://step.esa.int/main/ download/snap-download/
- Science Toolbox Exploitation Platform forum (https://forum.step. esa.int/)

6.18 Key Terms

- Copernicus program: Series of Earth Observation missions from the European Union, which offer free-to-access satellite data.
- Digital terrain model: Topographical model of the Earth used to help correct satellite images for variations in the terrain.
- Ground control points: Known latitude and longitude points used to help calibrate the geometry within satellite images.
- Normalized Difference Vegetation Index: Satellite algorithm that's used to show how green an area is, by highlighting its green vegetation.
- Sentinel-2: A twin optical satellite mission from the Copernicus program.

- Sentinel Application Platform: The suite of software used by this book to undertake practical image processing.
- Universal Transverse Mercator: A map projection used to allow maps of the Earth to be printed onto a flat 2D surface.

7

Geographic Information System and an Introduction to QGIS

The previous two chapters have focused on image processing techniques; however, remote sensing is more than just image processing. Remote sensing data can create maps, which become particularly powerful when combined with other geographical information. Image processing packages, like the Sentinel Application Platform (SNAP), don't have the full functionality to work in this manner. Therefore, it's necessary also to use a Geographic Information System (GIS) to complement the image processing software.

This chapter will outline the principles of a GIS, introduce the package Quantum GIS (QGIS), and give some practical exercises using remote sensing data within QGIS. This will complete the foundation for the book's second half, where the practicals will use QGIS and/or SNAP.

7.1 Introduction to GIS

The term *GIS* was first used by Dr. Roger Tomlinson (1968), while working for the Department of Forestry and Rural Development in Canada, in his paper "A Geographic Information System for Regional Planning". However, it wasn't until the latter part of the 20th century, and the advent of personal computing, that GIS began to develop a strong user community. Over the years, GIS has been defined in several different ways, with authors (such as Burrough and McDonnell, 1998) primarily describing it as a powerful set of tools that take collected and stored data and transform it into displayed spatial data, that is, data with geographical (or xy) coordinates.

As technology and computing power have developed, the focus has moved from simply displaying data to practical applications. As a result, planners, decision-makers, and geographers working on a wide range of subjects now commonly use GIS software. The essence of GIS software is the multiple ways it can handle data. It uses both the raster layers (as described in Section 5.1) and vector layers that include points, lines, and polygons:

DOI: 10.1201/9781003272274-7

- Points only possess a single *xy* coordinate, and their location is depicted on an image by a small symbol, such as a filled circle. Layers can be created for distinct features such as cities, airports, streetlights, individual trees, or other points of interest.

- Lines are a sequence of at least two pairs of *xy* coordinates, which define a straight line, and several pairs of *xy* coordinates for curved or complex-shaped lines. Lines are used for features such as the centerlines of rivers, streets, or elevation contours.

- Polygons are features with boundaries that have a sequence of *xy* coordinates where the starting point is also the endpoint. This approach creates a two-dimensional (2D) shape feature, with a calculable area, used to represent lakes, buildings, and even countries.

As the data in a GIS come from various sources, they require a consistent framework for pinpointing the locations of features in the real world, known as the Coordinate Reference System (CRS). In general, CRS combines two elements: a projection and a datum.

The problem with displaying maps of a three-dimensional (3D) Earth, onto 2D paper or computer screens, was introduced in Section 6.4 in relation to displaying Landsat scenes. When 3D data are flattened onto a 2D projection, not all the properties can be retained; essentially, it's only possible to have either correct angles, correct areas, or correct distances.

- Correct angles: The most common projection for global data is the geographical or zero projection, referred to as "World Geodetic System of 1984 (WGS84)," or similar, within GIS systems. This is a projection with parallel lines of latitude and longitude, and an example was shown in Figure 2.1b. It retains the correct angles but shows incorrect areas, and distance cannot be easily measured. It's referred to as the World Geodetic System of 1984 (WGS84) in QGIS and "Geographic Lat/Lon (WGS84)" or "EPSG:4327—WGS 84 (geographic 3D)" in SNAP.

- Correct area: If the data are reprojected onto an equal area-focused global projection, such as the Mollweide projection shown in Figure 7.1, then the areas of the landmasses at the high latitudes will become much smaller to reflect their actual size, but the angles and distances are distorted. This type of projection is helpful if you're planning to calculate areas over large geographical regions.

- Correct distances: The Universal Transverse Mercator (UTM) projection used by Landsat, as was described in Section 6.4, has correct distances, enabling the size of a pixel to be defined in meters.

FIGURE 7.1
MODIS global data reprojected into a Mollweide Equal Area Projection. (Data courtesy of USGS/NASA.)

The area calculation is reasonable for small areas but has increasing errors as the area covered becomes bigger, and thus the UTM projection divides the Earth up into several thin strips to keep distortions to a minimum.

The second element of the CRS is the datum, which provides the framework for measuring distances and heights on the surface of the Earth. A common height datum you may be familiar with is sea level, where locations are referred to as above or below sea level; however, there is no worldwide agreement on what sea level height actually is, and hence it varies between countries, meaning it cannot be easily used within a GIS. The most common datum used for satellite data sets is the one that's part of WGS84; it uses an origin located at the center of the Earth to measure heights above a global oblate spheroid and is believed to be accurate to within 0.02 m. It is commonly used for GIS data sets because it can be used for the whole Earth, but it has limitations as discussed above.

Projection difficulties can be reduced by displaying satellite data on virtual globes, which is increasingly occurring, with Google Earth being the best-known example. However, even this isn't perfect, as it assumes that the Earth is a sphere rather than its actual shape: a squashed sphere that's flattened at the poles and bulges at the equator, also called an "oblate spheroid". Coupling this with the Earth's irregular surface means that virtual globes still have inbuilt CRS errors.

7.2 GIS Software Packages

As with image processing packages, various commercial and free open-source GIS solutions are available. ArcGIS, offered by Esri, is probably the best-known commercial solution, although there are others, such as GeoMedia by Hexagon and MapInfo by Precisely. There are also several open-source GIS solutions available, including Geographic Resources Analysis Support System (GRASS, https://grass.osgeo.org/), originally developed by the US Army Construction Engineering Research Laboratories, and the System for Automated Geoscientific Analyses (SAGA) originally developed by J. Boner and O. Conrad at the Institute of Geography at the University of Gottingen; both of which are included within QGIS.

Continuing the approach taken with the remote sensing package in this book, we've chosen to use QGIS as the demonstration GIS package; it's a free open-source solution that enables the creation, editing, analyzing, and visualizing of geographical information, and it's actually a collection of packages, including GRASS and SAGA, making it particularly powerful software. In addition, it has been around for several years and has a sizeable international community supporting its development with regular updates and multilanguage support. Similarly to the image processing section, although we're using QGIS, the principles discussed and the associated learning should be transferable to other GIS packages.

QGIS provides a set of web pages as an introduction to using GIS, which provides further background information on a number of aspects, such as the CRS as discussed in Section 7.1 (https://docs.qgis.org/3.22/en/docs/gentle_gis_introduction/index.html/).

7.3 Installing QGIS

From the QGIS web page (https://www.qgis.org/), you should see a Download Now button that takes you to another page offering the choice of download version depending on the computer's operating system. If you're using Windows, choose the QGIS Standalone Installer Long Term Release version for new users. The latest Long Term Release version at the time of writing is Bialowieza 3.22, which will be the basis of the examples in this chapter. QGIS is only available as a 64-bit version. If you download a more recent version of the software than we use in the book, it might mean that the look, menus, and options may be slightly different. An online list of resources is available via the learning resource website (https://playingwithrsdata.com/), which will be updated to support any software developments linked to the practicals.

Selecting the hyperlink for the appropriate software version for your system sets off a standard installation approach. Once downloaded, the installation may run automatically, or you may need to open the downloaded file to start the installation. It's worth noting that, like image processing software, GIS software requires a large amount of disk space. The current QGIS compressed installation file is 1 GB, and the final installation requires approximately 2.3 GB; therefore, you'll need to ensure there is sufficient free disk space for both the package and remote sensing data you'll be using. Like SNAP, QGIS is installed through an installation wizard; you don't need to make any changes to the default options it offers during the installation, but it does take a little while to install.

Once QGIS is installed, you'll see a folder-like icon on your desktop titled QGIS 3.22; there may be an additional number to indicate which version of the Long Term Release it is. If you Double-click this icon, it will open a selection of shortcuts:

- QGIS Desktop 3.22 is the main software that this handbook focuses on.
- GRASS GIS gives access to this specific GIS package.
- SAGA GIS gives access to this specific GIS package.
- OSGeo4W bundles several geospatial tools together and allows a customized setup. The Setup tool allows you to set these up while Shell provides a command prompt, rather than a graphical user interface. To use these is a step beyond what's covered in this handbook, so we suggest you review these online if you want to customize your setup.
- Qt Designer with QGIS 3.22.7 custom widget is a tool that allows designers to develop geographical interfaces for QGIS plugins.

For this book, you'll only need to use the QGIS Desktop. So, select this, right-click on your mouse, and select Send to > Desktop, and you'll send the main QGIS icon to your desktop. You can now just double-click this icon on the desktop to open QGIS ready to use.

NOTE: If you experience problems with the QGIS Toolbox, then check the directory structure your files are in. Ideally, this should be as short as possible, and directory/filenames should not have spaces as this can cause problems. Questions can be asked on the Geographic Information Systems Stack Exchange (https://gis.stackexchange.com/tour).

7.4 Introduction to QGIS

On the first occasion, the QGIS Desktop is opened, it may take some time because, like SNAP, it's composed of many individual components that need to load. However, it will be quicker when it's opened on successive occasions, as it will have remembered your layout preferences and last-used tools.

QGIS presents a similar-looking graphical user interface to SNAP. The menu bar with toolbars is running along the top of the screen. The main window is where the images and data sets are displayed. Running along the bottom of the screen is the status bar showing details about the current data sets, which indicates the current map Coordinates, Scale, Magnification, Rotation, and map projection. Also, a blue processing bar appears here when QGIS is processing, which is helpful as it tells you it is still working on your last action. As QGIS deals with more than just raster images, what's displayed is named as layers rather than images.

Next to the left toolbar are a series of windows, referred to as the "Layers/Browser Panel". You'll see the default options the first time QGIS is opened:

- Browser: A version of Windows File Manager within QGIS, which can be used to access data files quickly
- Layers: The list of what you're currently working on in QGIS and is approximately equivalent to the Product Window in SNAP

The main window is the Map Display/Recent Project and shows the data you are currently working on. You will also see the News feed when you initially open QGIS, which will disappear when you load data or a recent project.

The Layer and Browser panel windows can be moved around and closed, like any window. If you right-click over the name of any window, you'll get a list of potential options enabling personalization of this screen area to the elements most frequently used. **NOTE:** If you have too many panels open, they will disappear below the level of your screen and become inaccessible.

Finally, it's worth highlighting that QGIS should periodically have its plugins updated. This update is through the menu item Plugins > Manage and Install Plugins. If any upgrades are necessary, you can click the Upgrade All button. This approach will ensure you'll have the most up-to-date version of the software, and known bugs removed. The Upgrade All button will be grayed out if there are no available upgrades.

We'll move into the practical using QGIS in the next section, and we'll take you through each element step-by-step. However, as we will only cover a small part of QGIS in the book, it is worth knowing that there is an online QGIS User Guide to give more details on the package (https://docs.qgis.org/3.22/en/docs/user_manual/index.html).

7.5 Importing Remote Sensing Data into QGIS

For this practical, we will use the same Landsat imagery for Poland that was used in Chapter 6 for SNAP, that is, the file "LT05_LT1P_1880 24_19901027_20200915_02_T1.tar" so you should already have a folder containing the extracted Geostationary Earth Orbit Tagged Image File Format (GeoTIFF) files. If not, download it again using the instructions in Section 6.6.

QGIS can't use the MTL file to import data; instead, it imports GeoTIFF files directly. There are several ways you can do this: importing the spectral wavebands individually or creating a single GeoTIFF file containing all or several wavebands imported into QGIS. We'll look at the different ways of importing the data into QGIS via other chapters. However, for this chapter, we will create a single GeoTIFF containing all the spectral wavebands.

We're going to create this single GeoTIFF file within SNAP. To do this, load the MTL file into SNAP as you did in Section 6.7, click on Bands, then go to the menu item File > Export > GeoTIFF/BigTIFF. This approach will give a file-saving dialog box and create a file called "LT05_LT1P_188024_19901027_20200915_02_T1.tif". **NOTE:** This export may take a short time.

This file can then be imported into QGIS using the menu item Layer > Add Layer > Add Raster Layer and then click on the ... at the end of the Source box. This opens a File Explorer window to allow you to navigate to the appropriate file and click Open. You are now back at the Data Source Manager Raster window and can click Add, which will import the GeoTIFF. The Data Source Manager Raster window will remain open, but under the Layers panel, you should now see the name of the file and some of the bands. Click Close to shut the Data Source Manager Raster window. Alternatively, to get to the Data Source Raster Manager use the first icon on the second row; the Open Data Source Manager icon.

Once the data have been imported, they will be listed in the Layers panel – if you can only see a small version of the main image, close down any other open windows on the right side of the screen, so you have a view like Figure 7.1. The main view will be the imported Landsat scene, but it will have a red hue and may not look as you might have expected.

This hue is because QGIS has automatically chosen the spectral wavebands to display and their order. To change the displayed wavebands, you need to amend the properties of the image. These properties can be accessed either by selecting the layer and then going to the menu item Layer > Layer Properties, or by right-clicking on the layer name in the Layers panel and choosing Properties. This will open a dialog box with a number of tab options, with the default being the Symbology tab; the window is slightly bigger than the default box, so you may have to scroll left and right to read the window.

At the top of the Symbology tab is the "Band Rendering" options, and you'll see the spectral wavebands automatically chosen by QGIS. Initially, it picks wavebands 1, 2, and 3 for the red, green, and blue bands, respectively, as shown in Figure 7.2a. From working on RGB images in Chapters 5 and 6, you'll know that to create a pseudo-true-color composite for this Landsat Thematic Mapper (TM) scene the RGB image needs wavebands 3, 2, and 1 selected in that order. Use the drop-down menus for each waveband to swap the red and blue wavebands. The image will now look like Figure 7.2b. **NOTE:** It can take a minute or two for QGIS to complete this processing; depending on the speed of your computer, you'll see the blue processing bar appear, and while this is onscreen, you know QGIS is still working on the processing.

As with SNAP, in QGIS, you can zoom in and out of an image, plus pan around the image using the first row of icons at the top of the screen. The hand icon is for panning, and there are multiple magnifier icons to zoom in various ways.

7.6 GIS Data Handling Technique: Contrast Enhancement/Histogram Stretch

In Section 7.5, the default settings were used for the histogram stretch that was "Stretch to MinMax". However, the image can be improved by applying a custom stretch to enhance the optimum contrast. Unlike SNAP, within QGIS, the histogram has to be calculated; to do this, select the Histogram tab from the Layer Properties and click Compute Histogram. You'll now see all the wavebands displayed together on the histogram. To create a custom stretch, you'll need to manually enter the minimum and maximum values for each waveband.

Adjusting the range can be done under the Histogram tab by changing the wavebands using the drop-down menu at the bottom or by entering the values in the relevant boxes on the Symbology tab. However, although you changed the order the bands are displayed in Section 7.5, within

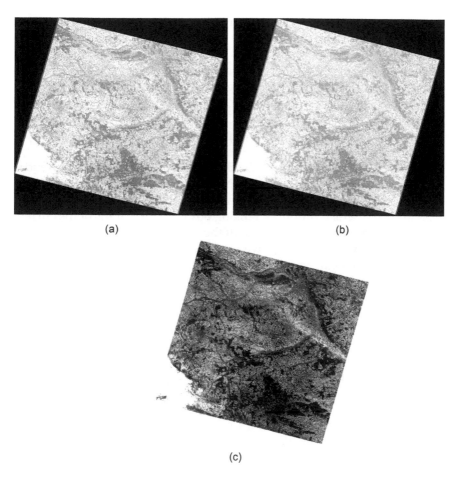

(a)　　　　　　　　　　　　　　　　(b)

(c)

FIGURE 7.2
Landsat-5 Thematic Mapper scene over Europe (path 188, row 024) displayed as (a) view as immediately imported into QGIS, (b) pseudo-true-color composite with (c) an improved contrast enhancement stretch applied to improve what's visible. (Data courtesy of NASA/USGS.)

the Histogram the wavebands are still in the original order they were imported, that is, Band 1 is Blue, Band 2 is Green, and Band 3 is Red.

The histogram will be fairly small on the default graph, so it can be useful to zoom in. If the mouse is moved onto the histogram itself, a magnifier will appear, and as you hold down your left mouse button, a box can be drawn around the area of the histogram you want to zoom into. Pressing the right mouse button will return the histogram to the default view. All the wavebands are shown on the default graph, but if you click the downward pointing triangle beside Prefs/Actions, you'll have several options, and select "Show RGB/Gray band(s)" as now only those bands will be on the histogram. The minimum and maximum values for the histogram stretch we've chosen are 50–70 for Band 1, 20–30 for Bands 2 and 3.

Finally, go to the Transparency tab, and set the "Additional no data value" property to zero. This change will remove the more prominent black square surrounding the image. Click Okay, and all the changes will be applied to the view. **NOTE:** The Okay button applies the changes and closes the dialog box, while Apply applies the changes but keeps the dialog box open.

The result, as seen in Figure 7.2c, shows a similar image to the pseudo-true color RGB image produced in SNAP. The view can be exported from QGIS by selecting the menu item Project Menu > Import/Export> Export Map to Image, which produces a dialog box with the details of the current image in the map window – although there are several properties you can change if you want. Clicking Save produces a standard file-saving dialog box, and the view can be saved in various formats.

7.7 GIS Data Handling Technique: Combining Images

As you've already seen, when downloading Landsat data, you get an individual image of a specific geographical extent. This extent may not be big enough for your needs, or the features you're interested in may cross multiple images. Within QGIS, it's possible to combine multiple images together very accurately, because the Landsat scenes are georeferenced, and the CRS used is compatible. When two images are loaded into separate raster layers of the GIS, one layer will overlay the other. With raster layers, it's only possible to see the top layer, unless it's made transparent, enabling the layers underneath to become visible. The Layers panel indicates the order; that is, the one at the top of the list will be the layer on top in the viewer. The X in the checkboxes next to the layer names indicates the layers displayed, and by clicking on the X, it's possible to switch off a layer and the X disappears from next to its name.

To demonstrate the process of combining images, a second Landsat scene must be downloaded that overlaps geographically with the current scene. We have chosen a Landsat-4 TM (rather than a Landsat-5 TM) scene; it has Worldwide Reference System path 187 and row 24 and was acquired on May 15, 1988. Go to GloVis, and download the Landsat collection Level-1 Product Bundle for this scene, which will provide a tar file of approximately 247 MB in size with the filename "LT04_L1TP_187024_1988 0515_20200917_02_T1.tar."

Following the same process as in Section 7.5; namely, unzip the file, import the Landsat-4 TM scene into SNAP, create the single file of all wavebands, import this file into QGIS by adding it as a raster layer, change the displayed wavebands, compute the histogram, optimize the histogram stretch, and set the additional data value to zero. The histogram stretch values we've used for the new Landsat-4 TM image are 70 to 120 for Band 1, 25–60 for Band 2, and 20–80 for Band 3.

You will now have the two Landsat scenes accurately overlapping to show a larger area of Poland. In the Layer panel, the Landsat-4 TM scene will be the top layer as it was the last one imported; where the scenes overlap, it is the Landsat-4 TM scene that's visible. This can easily be altered in the Layer panel by selecting and dragging the Landsat-4 TM layer below the Landsat-5 TM one. In the Layers panel, the layer order should be reversed, and the scenes displayed will match Figure 7.3a. Zooming in on where the two layers overlap, such as in Figure 7.3b, demonstrates how accurately the scenes align. You'll notice that the colors within the two scenes are not identical; it's very difficult to get two L1 scenes taken on different days and by different missions to match precisely because the sensor calibration and atmospheric conditions will vary. The match can be improved by applying an atmospheric correction, as explained in Section 9.5.

7.7.1 GIS Data Handling Technique: Combining Data from Different Satellites

We've just combined two Landsat scenes, but now we're going to go one step further and add a Sentinel-2 image to the two Landsat scenes. Combining multiple satellite images gives greater flexibility in finding cloud-free images over the area at the time you are interested in.

The first step is to download the image from GloVis. Use latitude and longitude of 52.15 N and 21.00 E, respectively, both positive. Select Sentinel-2 from the Data Set list. Put in the Common Metadata filters for the data range of January 01, 2021 to March 31, 2021, and a cloud cover of 0–30. This should take you to Warsaw and offer 19 scenes matching these criteria. You'll see several partial images for this area, as you'll sign the jagged black line on the main image, and the mostly black tiles in the thumbnails. Scroll through, and see the different images available, for this exercise we're selecting the image from February 06, 2021, and although there are some clouds down the bottom of the image, over Warsaw it is clear. Download the JPEG2000 option from GloVis will give you an approximate 800 MB zip file called LC1_T34UDC_A029391_20210206T095357. If you unzip this file, you will get the folder S2A_MSIL1C_20210206T095201_N0209_R079_T34UCD_20210206T110416.SAFE with all the data inside.

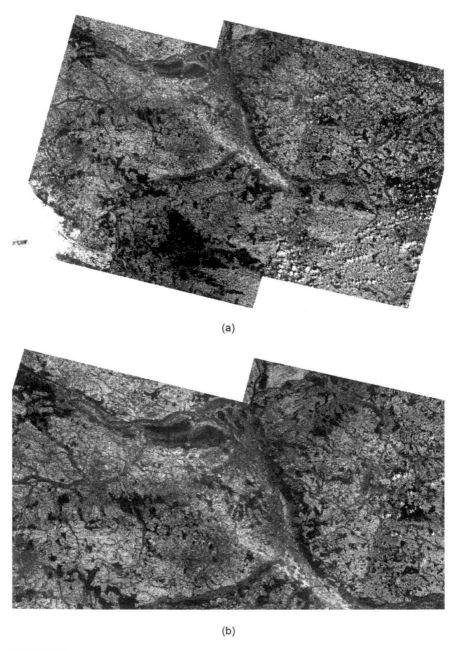

(a)

(b)

FIGURE 7.3
Combined Landsat-4 and Landsat-5 TM scenes displayed as (a) the full extent, and (b) zoomed-in overlapping area. (Data courtesy of NASA/USGS.)

Follow the steps in Section 6.15.3 to import the Sentinel-2 data into SNAP using the MTD_MSIL1C file. If you now attempt to create a GeoTIFF/BigTIFF, as you did for the Landsat images earlier in this chapter, you'll get an error message saying, *"Writing this product as GeoTIFF-BigTIFF is not possible: Cannot write multisize products. Consider resampling the product first"*. This is because, within the Sentinel-2 image, there are wavebands with different pixel sizes, so the first step is to resample the image so that all the wavebands have the same pixel size. Select Bands and then go to the menu item Raster > Geometric > Resampling, a dialog box will appear with I/O Parameters and Resampling parameters tabs. Go to the Resampling parameters tab, where you can choose how you want to resample. For this exercise, we are going to resample on Band, currently, it will show B1 with a width and height of 1,830. If you press the downward pointing triangle at the end of B1, and select B2, you'll notice the height and width will increase to 10,980. Press Run, and a new product will appear in the Product Explorer window with suffix_resampled.

Practically, you can now export this as GeoTIFF/BigTIFF. To do this, select the new resampled image, click on Bands and go to File > Export> GeoTIFF/BigTIFF, and get the export dialog box up, followed by clicking on the Subset button; over on the right side, which brings up the various ways you can subset images. As we want to overlay this image onto the two Landsat images, we don't want to subset the image spatially as we previously did. This time we are going to use the Band Subset instead. Go to the Band Subset tab, and you'll see all 13 bands of Sentinel-2 are selected, and notice that the estimated storage size of the image of around 16GB! As Sentinel-2 is a much bigger image than Landsat, due to the 10 m versus 30 m spatial resolution, this is going to be too big to handle. As we are looking to create an RBG image, it is better to create the GeoTIFF/BigTIFF using only the bands needed for the RBG image, which are Bands 2, 3, and 4. So click the checkbox to "Select none", which removes all the bands, and then click on the checkboxes to add B2, B3, and B4. With these three selected, the estimated storage size is a more manageable at 690 MB, so click Okay. Change the location of the saved file if you need to, and now click "Export Product". Even with the subsetting, this processing will take some time, so now is a great opportunity to take a break while completing this action.

Once the GeoTIFF/BigTIFF has been exported from SNAP, it can be imported in the same way as for Landsat, by going to Layer > Add Layer > Add Raster Layer, and selecting the resampled Sentinel-2 file you've just exported. The Sentinel-2 layer will straddle the two Landsat layers, as the Sentinel-2 image is the last one imported, it will be on top. QGIS better recognizes Sentinel-2 data and will have already imported the bands in the correct order. Go to the Histogram stretch tab, Compute Histogram, and then zoom in. As you only exported the RGB bands on the Sentinel-2 image, these are all that are shown on the histogram. Previously, we've entered the Min and Max values directly. Still, an alternative approach is putting the cursor in the Min box, clicking on the Pointing Finger symbol next to the box, and then you can click on the histogram to select the value you want. You can repeat this for the Max value and all the individual bands. For this image, we've chosen values of 1,100 and 2,000 for the Red band, 750 and 1,750 for Green, and finally 500 and 2,000 for the Blue band. Also, add the Additional No Data Value of 0 into the transparency tab. Click Apply, then Close.

You'll now have three layers all overlapping, similar to Figure 7.4a. If you zoom in to the border of the Sentinel-2 and Landsat layers, you'll see they line up with each other, but also the increased spatial resolution of Sentinel-2 is clear as the image is more detailed and sharper than the Landsat image; as can be seen in Figure 7.4b.

It's possible to add more layers to create larger images, but be aware that increasing the number of layers requires greater computing power. Adding too many layers may slow a computer down and increase the processing and waiting time to display the layers.

7.8 GIS Data Handling Techniques: Adding Cartographic Layers

It's also possible to download freely available cartographic vector layers and raster layers to add more detail and features to your view. There are several places where you can find these layers, such as Esri ArcGIS Hub (https://opendata.arcgis.com/) and United Nations Environment Programme (UNEP)/GRID-Geneva metadata catalog (https://datacore-gn.unepgrid.ch/geonetwork//srv/eng/catalog.search#/home). For the next part of the exercise, we will use the Natural Earth public domain map data set (https://www.naturalearthdata.com/), which the North American Cartographic Information Society supports, as it is free and has a wide range of available layers.

Go to the Natural Earth website and then select the Downloads tab, which offers three levels of data (large, medium, and small scale). Within each is a series of potential layers to download:

- Cultural layers include vector data for countries and sovereign states, population centers, airports, ports, parks, and time zones.
- Physical layers include vector data for coastlines, landmasses, oceans, rivers, lakes, reefs, islands, glaciers, playas, major physical features, and geographic lines.

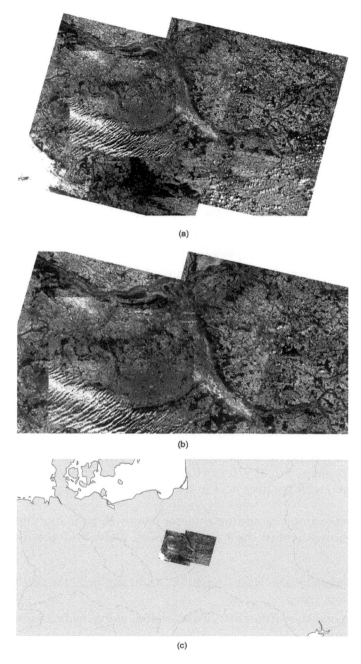

FIGURE 7.4
Combined Landsat-4 and Landsat-5 TM scenes plus Sentinl-2 scene displayed as (a) the full extent, (b) a zoomed-in overlapping area overlain with the Natural Earth 110-m Populated Places layer and 50-m Airports layer, and (c) the Natural Earth 50-m land layer with the river and lake centerlines. (Data courtesy of NASA/USGS.)

- Raster layers include raster data for bathymetry data, elevation contours, and the ocean bottom.

Each of the groups of layers can be downloaded in total, or as individual layers, and are downloaded as a zipped file containing many individual files. You'll have a series of layers offered if you click on the Cultural link under the Small Scale Data. Scroll down to the Populated Places layer, and click on the Download Simple (fewer columns) option. This will initiate the download of a zip file, and once extracted, there will be seven files:

- ne_110m_populated_places_simple.cpg
- ne_110m_populated_places_simple.dbf
- ne_110m_populated_places_simple.prj
- ne_110m_populated_places_simple.README.html
- ne_110m_populated_places_simple.shp
- ne_110m_populated_places_simple.shx
- ne_110m_populated_places_simple.VERSION.txt

These vector data are held in a shapefile, although it comes as a minimum data set of three files. The .shp file is the main one storing the vector coordinate data, the .shx file holds the index of data in the vector file. The .dbf file contains the attribute data, and .cpg file contains data to specify the encoding in the .dbf file. The .prj file includes the CRS definition, while the.html and .txt files are the website link and the downloaded data version, respectively.

If you download the raster data, the files are slightly different; this zip file only contains five files. It has the same .prj, .html, and .txt files as the vector data, with the data held in the .tif file and its accompanying .tfw CRS file.

The first example of adding a vector layer will be to highlight a couple of features within the Landsat scenes. Download the two vector layers:

- Populated Places layer from the cultural section of the small-scale 1:110 m group
- Airports layer from the cultural section of the medium-scale 1:50 m group

Once all the files have been extracted, they can easily be imported into QGIS. Go to the menu item Layer Menu > Add Layer > Add Vector

Layer and, from the dialog box, use the ... to locate the shapefile for the "ne_110m_populated_places_simple.shp" layer. Then, select the layer by pressing Open and then Add to import it. You'll see a new layer at the top of the layer panel for ne_110m_populated_places. Repeat this approach to import the "ne_50m_airports.shp" file.

If you look closely at the view, you'll see two dots in the top right corner of Sentinel-2 layer, indicating a populated place and an airport. It's possible to add a label to each layer by clicking on it. In this case, start by selecting the ne_50m_airports layer, then select the menu item Layer > Layer Properties followed by the Labels Tab. Alternatively, right-click on the layer and select properties or press the abc button, within the second row of icons. All these approaches will open up the Labels dialog box where there will be a property No Labels. Click on the downward-facing triangle next to this, select the option for Single Labels, and then change the label color from Black to White to make them easier to see and click Apply. Repeat these steps with the ne_110_populated_places_simple layer. The two dots will now be marked Warsaw and Okecie Int, respectively, as shown in Figure 7.4b.

Before undertaking a second example, we'll first remove from view the Populated Places and Airport layers. This change can quickly be done by clicking the checkbox, which will remove the tick from the box, and the layer will no longer be visible. Clicking the checkbox again will make the layer visible again.

Next, download both the Land, and the Rivers and Lakes Centerline layers from the medium-scale 1:50 m Physical section. Once extracted, import the shapefiles "ne_50m_land.shp" and "ne_50m_rivers_lake_centerlines.shp". Arrange these layers so that the rivers layer is at the top and the land layer is at the bottom with the three satellite layers between; left-click on the layer name, and drag it with the mouse to move the layers. Adjust the color of the land layer by selecting the Properties Symbology tab and using the drop-down menu to change the color. Click the downward pointing triangle at the end of Color, and from the menu, click Choose Color, which will bring up the color options. Go to the B option on the right side by selecting that radio button. You should now have a much lighter color palette in front, so select one of the lighter green color shades in the bottom right corner of the palette. Click OK, and then OK again when you are back at the Layer Properties tab. Zoom out of your image a few times you'll see something similar to Figure 7.4c, where the three satellite images have been set within the European landmass with the coastline and the ocean visible. The main rivers of the area can also be seen with the Vistula River running through Poland and the two Landsat images.

7.9 Coordinate Reference System Adjustments within QGIS

At the start of this chapter, in Section 7.1, the CRS system was explained, and it was noted that the most common CRS used by GIS data sets is WGS 84. If you go to Properties on one of the vector layers and select the General tab, the CRS used is noted; in this case, it is "EPSG:4326, WGS 84", indicating that the layer uses WGS 84 for both datum and projection. Repeating this action for one of the Landsat layers produces a different CRS, namely, "EPSG:32634, WGS 84/UTM zone 34 N", indicating that Landsat uses WGS 84 for datum but is using UTM zone 34 N for the projection.

QGIS is automatically reprojecting the cartographic vector data to Landsat's UTM projection so that the different layers correctly overlay each other. The EPSG Codes are short-hand for defining CRS and are set out by the International Association of Oil and Gas Producers (https://epsg.org/home.html).

If you change the CRS used by a layer, then it will move the data within the GIS. This change is because the position will be relative to a different reference point, even though the coordinates will be the same. Many users find this difficult to understand, and even experienced users can get it wrong, leading to situations where organizations have searched for resources such as oil and built buildings in the wrong place. Hence, be very careful when changing the CRS used by a layer.

7.10 Saving Images and Projects in QGIS

Images can be saved in QGIS, similar to SNAP; go to the menu item Project > Save As Image, and you'll be given a standard file-saving dialog box with a default.png file, although this can be changed to other image formats. It takes a short period to export the image, but the processing can be seen in the status bar at the bottom as the blue processing indicator, and when completed, a green bar at the top of the main image will confirm that it has been exported.

Alternatively, if you want to save the whole QGIS project, to continue working at a later date, use the menu item Project > Save As, and the standard file save dialog box enables the QGIS project to be saved in a .qgis format.

7.11 Summary

Manipulating remote sensing data effectively requires both image processing and GIS software. This chapter has introduced you to the second type through QGIS and has gone through some basic data handling techniques. These techniques include loading and viewing raster and vector data, manipulating raster layer colors through histogram stretches, and displaying the attributes of vector layers as an annotation. Knowing this provides the equivalent basic GIS skills to those you've already learned for image processing.

You've now experienced both computer packages we will use in this book. The remaining practical exercises include either, or sometimes both, of these packages, particularly highlighting how they can be used for real-world applications.

Like SNAP, the best way of learning QGIS is to plunge straight in and start to experiment; the more you use these packages, the more confident you'll become in applying remote sensing techniques. Go on, give it a go!

7.12 Online Resources

- Associated learning resource website: https://playingwithrsdata.com/
- EPSG Geodetic Parameter Data set: https://epsg.org/home.html
- Esri ArcGIS Hub: https://opendata.arcgis.com/
- Geographic Information Systems Stack Exchange, forum for QGIS queries: https://gis.stackexchange.com/?tags=qgis
- GRASS GIS: https://grass.osgeo.org/
- Natural Earth: https://www.naturalearthdata.com/
- QGIS A Gentle Introduction to GIS: https://docs.qgis.org/3.22/en/docs/gentle_gis_introduction/index.html
- QGIS main web page: https://www.qgis.org/en/site/
- QGIS User Guide: https://docs.qgis.org/3.22/en/docs/user_manual/index.html
- United Nations Environment Programme (UNEP)/GRID-Geneva: https://datacore-gn.unepgrid.ch/geonetwork//srv/eng/catalog.search#/home.

7.13 Key Terms

- Coordinate Reference System or Projection: Defines how a 2D projected map is related to real locations on a 3D Earth.
- Datum: Reference used within a GIS for measuring distance and heights on the surface of the Earth.
- Geographic Information System: Set of tools that take collected data and display it in a spatial format using geographical xy coordinates.
- Vector data: Such data have xy coordinates and can be either individual points or groups of points to form lines or polygons.

References

Burrough, P. A. and R. A. McDonnell. 1998. *Principles of GIS*. London: Oxford University Press.

Tomlinson, R. F. 1968. A Geographical Information System for rural planning, Esri the 50th Anniversary of GIS. Available at https://gisandscience.files.word-press.com/2012/08/1-a-gis-for-regional-planning_ed.pdf (accessed April 17, 2015).

8

Urban Environments and Their Signatures

8.1 Introduction to Application Chapters of the Book

This chapter is the first in the second half of the book, which focuses on applications, providing an introduction to urban, landscape, inland water, coastal, and atmospheric remote sensing. It gives an insight into what's possible, although, within this book, we can only scratch the surface of the applications. Chapter 13 will describe where you can go to find more information and progress to the next stage of your learning.

Each of these chapters will be organized in the same way; that is, they'll start with an introduction to the area, followed by the remote sensing theory, then the individual applications, and we finish the chapter with a practical exercise using Quantum Geographic Information System (QGIS) and/or Sentinel Application Platform (SNAP) together with readily available data sets.

8.2 Urban Environments

The urban environment can be described as the area surrounding and within a city, together with the associated living accommodation, shops, office/work buildings, transport infrastructure, and leisure facilities for the population of that city. These environments cover less than 3% of the Earth's land surface area, but more than 50% of the human population lives within them (McGranahan and Marcotuullio, 2005). Developing this concentration of facilities to support human populations changes the local natural biodiversity; some habitats are lost, while for others, the development offers new potential habitats that are scarce in the surrounding rural landscape. This diversity makes the urban environment a symbiotic ecosystem for both humans and other flora and fauna. This type of scenario is described as ecosystem services, which are the conditions and processes through which natural ecosystems and the species that compose them sustain and

DOI: 10.1201/9781003272274-8

fulfill human life (Daily, 2013); providing support for health and well-being together with both esthetic and physical influences. This support covers benefits such as food, clean air, flood management, climate regulation, healthy soils, renewable energy, and outdoor activities among others. The concept of ecosystem services was central to the Millennium Ecosystem Assessment (https://www.millenniumassessment.org/), published in 2005, highlighting that approximately 60% of urban ecosystem services have been degraded or used unsustainably.

Therefore, managing and maintaining ecosystem services to the benefit of all parties requires urban environments to be monitored over time, as they change and evolve. Satellite remote sensing can help this monitoring, by improving our understanding of how cities are laid out, including their relationships to adjoining areas, such as smaller settlements, and their impact on the environment in terms of factors such as air temperature, atmospheric gases, and particles, which is covered in more detail in Chapter 12.

This chapter begins with the theory of spectral and thermal signatures within urban environments, moving on to describe how remote sensing can be used to monitor specific urban environment applications, before focusing on an urban environment practical centered on analyzing Landsat and Moderate Resolution Imaging Spectroradiometer (MODIS) data within QGIS.

8.3 Introduction to the Optical Signatures of Urban Surfaces

Chapter 2 described how substances on Earth have their own individual spectral signatures. However, the remotely sensed signature does not remain constant as it's dependent not only on the composition and surface properties of the feature of interest but also on the geometric relationship between the angle of the incoming solar irradiance (solar zenith angle) and angle the sensor detects the signal from (sensor zenith angle). Figure 8.1 shows the sun on the left with the solar irradiance (E_d) being emitted toward the Earth, where E is used to represent irradiance and d signifies that the energy is coming downward from the sun; irradiance is the term for the energy received, known as flux, per unit area that's expressed using the SI units of watts per square meter, which will be shown in brackets when equations are written, that is, (Wm^{-2}).

As the electromagnetic (EM) energy from the sun passes through the atmosphere, it's scattered, resulting in energy being received in multiple directions, and some of the energy is absorbed by the Earth's atmosphere; as described in Section 2.3. This direction is noticeable when there is a cloudy day because, although there will be less light under clouds, it's not completely dark.

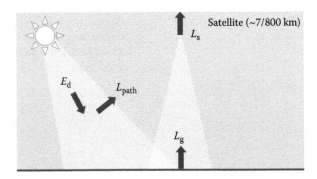

FIGURE 8.1
The path of electromagnetic radiation through the atmosphere, and interacting with the surface.

A proportion of the EM radiation that passes through the atmosphere and reaches the ground is reflected back toward space, which is called the ground radiance (L_g), and after absorption by the atmosphere, some of that radiation will eventually enter the sensor and is referred to as the sensor radiance (L_s). Radiance (L) is used once there is reflectance from the ground because we're now dealing with flux in a particular direction. Therefore, the SI units for radiance are watts per square meter per steradian (Wm^{-2}/sr), where the steradian represents a solid angle.

Having calculated the amount of solar irradiance the Earth is receiving from the sun, it is possible to determine the surface it strikes on the planet by measuring the amount of energy absorbed and reflected and then comparing these figures to the spectral library; Figure 8.2a shows examples from the ASTER spectral library (Baldridge et al., 2009) for urban surfaces.

In general, concrete is more reflective than asphalt, grass, and soil in the visible spectrum and, thus, will appear brighter in optical satellite imagery. Healthy green vegetation has a spectrum with strong features owing to the absorption of the pigments, primarily chlorophyll-a (Chlor-a), within the cells. The absorption within the visible and near-infrared (NIR) wavelengths, as seen in Figure 8.2b, shows a peak in the blue/green between 0.4 and 0.6 µm (i.e., 400 and 600 nm) that makes vegetation appear green.

Artificial features tend to have similar spectral signatures, as they're often composed of similar materials, except when they're being influenced by substances on their surface such as the accumulation of oil on a road, which changes the spectral signature and often lowers the reflectance. Also, there's the influence of shadows, which have their own spectral signature but can be broadly classified as being where the radiation is lower, so the pixels will be darker.

Therefore, optical signatures can be particularly useful in detecting elements when the feature of interest has a distinctly different reflectance

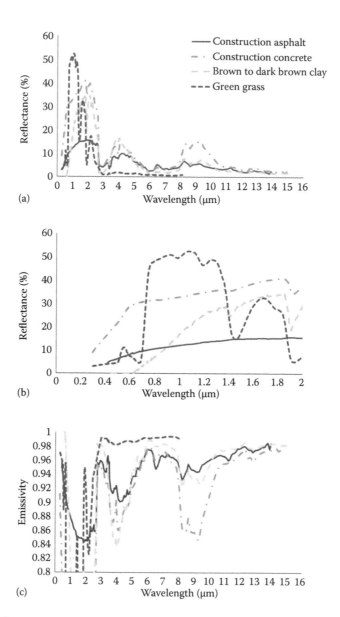

FIGURE 8.2
The ASTER spectral library for urban surfaces shown as reflectance for (a) a wide range of wavelengths and then (b) zoomed in for the visible and near-infrared (NIR) wavelengths with (c) showing the calculated emissivity spectra for the visible and NIR wavelengths. (Data courtesy of Baldridge et al. 2009. The ASTER spectral library version 2.0. *Remote Sens Environ* 113:711–715.)

spectral shape to the surrounding pixels. This spectral shape is capitalized on by waveband ratios and classification techniques, which will be explained further in Chapter 9.

8.4 Introduction to the Thermal Signatures of Urban Surfaces

While optical signatures are determined by measuring the reflectance of EM energy, thermal signatures measure the EM radiation emitted by the Earth. In remote sensing, this emitted radiation is mostly found within the far-infrared (IR) part of the spectrum between 8 and 14 μm; there's also emitted radiation in the mid-IR spectrum between 3.5 and 5.5 μm; however, this is more difficult to use during the day owing to reflected radiance and, hence, these wavelengths tend to be used for nighttime imagery.

The first step in measuring the thermal signature of an object is to determine the wavelength at which the body is radiating heat, which depends on its temperature. Wien's displacement law states the hotter an emitting body, the shorter the wavelength of the peak radiation, from the equation

$$\lambda_{max} = \frac{b}{T},\qquad(8.1)$$

where

- λ_{max} is the wavelength of peak radiation (m, which can be converted to μm)
- b is Wien's displacement constant $2.8977685 \pm 51 \times 10^{-3}$ (m K)
- T is the temperature in kelvin (K)

The Earth corresponds to a blackbody, a hypothetical perfect radiator of energy with no reflecting power, which has a temperature of around 300 K and, therefore, according to Equation 8.1, has a maximum peak of its emission at approximately 9.6 μm, within the far-IR spectrum. If there are heat sources, such as flames, then the peak will move to shorter wavelengths, and the mid-IR becomes important because this is now the location of the peak radiation.

While the Earth itself is a hypothetical blackbody, real materials aren't, and a conversion factor needs to be applied to calculate the amount of radiation the real materials emit, known as emissivity. It can be calculated using

$$\varepsilon = \frac{F_r}{F_b},\qquad(8.2)$$

where

- ε is the emissivity (unitless, expressed as a number between 0 and 1)
- F_r is the emitting radiant flux of the real material (Wm^{-2})
- F_b is the emitting radiant flux of the blackbody (Wm^{-2})

A perfect blackbody would have an emissivity of 1 and, thus, emits all the energy it has previously absorbed. Soil has an average emissivity of between 0.92 and 0.95, depending on whether it's dry or wet, while artificial materials such as concrete have an emissivity of around 0.937 (Wittich, 1997).

There is a second method of calculating emissivity, as it has an inverse relationship to reflectance. As the reflectance of materials can be measured in a laboratory, this method is often used as a practical way of calculating emissivity using Kirchhoff's law:

$$\varepsilon = 1 - R, \tag{8.3}$$

where

- ε is emissivity (unitless, expressed as a number between 0 and 1)
- 1 is the emissivity level of a perfect blackbody (unitless, expressed as the number 1)
- R is reflectance (unitless, expressed as a number between 0 and 1)

For example, Figure 8.2a shows the reflectance values of urban materials, soil, and green grass across multiple wavelengths; and, therefore, using Kirchhoff's law, the emissivity spectra, or thermal signatures, can be calculated as shown in Figure 8.2c.

Thermal signatures, like optical signatures, can be used to separate features where there are differences and can also be used to measure the temperature of features of interest such as land surface temperature (LST) over the land and sea surface temperature over the ocean.

8.5 Urban Applications

This section reviews a subset of urban applications where remotely sensed data can be used to aid understanding and knowledge.

8.5.1 Green Spaces and Urban Creep

Our towns and cities are increasingly changing, which brings added pressure on the environment. Development, especially that which displaces natural landscapes, can have unintended consequences; for example, building on, or paving over, vegetated areas can alter the capture and storage of rainwater and lead to an increased likelihood of flooding from the sewers and rivers, both locally and downstream. Therefore, tracking land cover changes over time is important for urban development planning and environmental resource management, with mapping impervious surface areas being a key indicator of global environmental change.

Satellite remote sensing can provide a good way of creating up-to-date land cover maps. The simplest methodology for viewing different land cover types is creating a visible/NIR false-color composite; for example, this could be done using Landsat Thematic Mapper (TM) data by displaying wavebands 4, 3, and 2 as red, green, and blue. Figure 8.3a shows an example for New York using Landsat ETM+ data; the vegetation appears red while the urban regions appear blue/gray.

Using a time series of these types of images can produce interesting snapshots of how local urban environments have changed and how this has affected the green space, indicating the changes likely from planned future developments.

8.5.2 Temperature Dynamics

Green spaces also play an important role in the quality of life, as they can significantly impact the local microclimate and the regional climate of a city (Hofmann et al., 2011). The replacement of natural surfaces with artificial ones can increase the urban heat island phenomena, where cities experience distinct climates because of variations in air temperature, humidity, and precipitation compared with their surrounding rural landscapes.

In addition, artificial materials such as concrete and tarmac can also reflect more radiation, and as these features can be relatively small, they offer greater diversity in the thermal radiation profile of an area. Therefore, by being able to distinguish these smaller spatial scales, thermal satellite imagery can provide a more spatially complete set of temperature measurements than in situ ones alone. This technique can also be used to measure heat loss from buildings to assess the efficiency and compare the heat usage of various factories or industries within urban environments.

Figure 8.3b shows the Landsat ETM+ thermal waveband for New York. It has been left black and white but stretched optimally to show temperature variations. Comparing this with the false-color-composite (Figure 8.3a), the cooler temperatures over vegetated versus artificial surfaces become apparent.

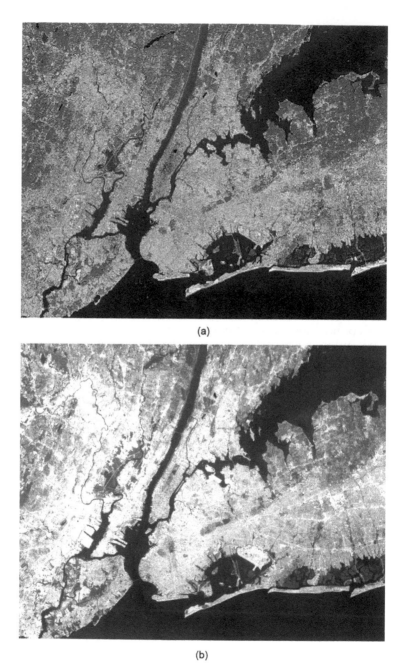

(a)

(b)

FIGURE 8.3
New York Bight shown using Landsat ETM+ data as (a) visible/NIR composite and (b) thermal waveband. (*Data courtesy of the U.S. Geological Survey.*)

If accurate values are required, such as for time-series analysis when looking at how a city has developed over time, then derived surface temperatures should be calculated through a correction for the effects of the atmosphere (which uses the emissivity, see Coll et al., 2010) and so the application of Planck's radiance function is

$$B_\lambda(T) = \frac{C_1}{\lambda^5 \left(e^{\frac{C_2}{\lambda T}} - 1 \right)}, \tag{8.4}$$

where

- $B_\lambda(T)$ is the derived brightness temperature (K) at a particular wavelength
- C_1 (Wm^{-2}) and C_2 (m K) are the first and second radiation constants
- λ is the wavelength (μm) for which the brightness temperature is being calculated
- $e^{\lambda T}$ is the wavelength multiplied by the temperature (K) as an exponential power

With Landsat TM and ETM+, this equation is transformed into the following simpler form:

$$T = \frac{K_2}{\ln\left(\frac{K_1}{L_\lambda} + 1 \right)}, \tag{8.5}$$

where

- T is the temperature (K)
- K_1 (Wm^{-2}/sr/μm) and K_2 (K) are sensor-dependent constants, as given in Table 8.1
- L_λ is the radiance (Wm^{-2}/sr/μm) at a specific wavelength, which is linked to K_1 and K_2
- ln is the natural logarithm

Derived surface temperatures can be calculated in image processing or GIS packages, using the mathematical module as demonstrated by applying a water index in Section 10.4 Step Five. Alternatively, Landsat products are routinely generated as Level 2 (L2) and can be downloaded from EarthExplorer; as used in the practical exercise in this chapter.

TABLE 8.1

TM, ETM+, TIRS and TIRS2 Thermal Band Calibration Constants

Sensor	K_1 (Wm^{-2}/sr/μm)	K_2 (K)
Landsat-4 TM	671.62	1,284.30
Landsat-5 TM	607.76	1,260.56
Landsat-7 ETM+	666.09	1,282.71
Landsat-8 TIRS waveband 10	777.89	1,321.08
Landsat-8 TIRS waveband 11	480.89	1,201.14
Landsat-9 TIRS2 waveband 10	799.03	1,329.24
Landsat-9 TIRS2 waveband 11	475.66	1,198.35

8.5.3 Nighttime Imagery

Light pollution is the brightening of the night sky caused by streetlights and other artificial sources; it disrupts ecosystems, wastes energy, and reduces the number of stars that can be seen. Such is the impact of light pollution that several areas around the world have been designated as dark-sky parks or reserves and are kept free from artificial light.

Since 1972, nighttime satellite imagery from the Operational Line Scan (OLS) system on the Defense Meteorological Satellite Program (DMSP) series of 24 weather satellites has been used to show urban light pollution (Elvidge et al., 2007). However, DMSP data were classified for many years, and digital data are only available from 1992.

The Visible Infrared Imaging Radiometer Suite (VIIRS) sensor, launched in 2011, now provides higher-quality imagery through the day–night band that collects panchromatic imagery by day and low-light imaging at night. It has a spatial resolution of 742 m, compared with 5 km for OLS, and a significantly improved dynamic range of 14 versus 6 bits, meaning a much greater range of brightness levels can be distinguished. In addition, the VIIRS measurements are fully calibrated, and thus quantitative measurements that allow for time-series comparisons are possible. VIIRS also collects nighttime shortwave infrared data to enable heat sources to be separated from light sources.

Figure 8.4 shows a VIIRS day–night band image from September 21, 2014, with Figure 8.4a showing southeastern Asia and Figure 8.4b zoomed in on the Thailand coast with numerous boats visible just outside the Bay of Bangkok in the Gulf of Thailand.

Also, high-resolution imagery is starting to be acquired by commercial satellites with the Jilin-1 constellation, built by Chang Guang Satellite Technology, including a satellite that acquires 0.92 m color video and nighttime imagery.

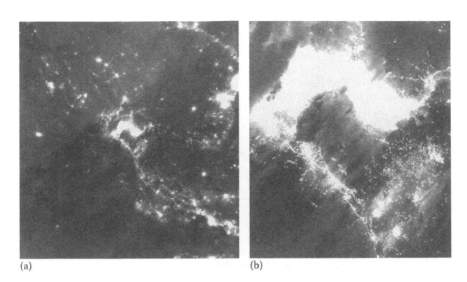

(a) (b)

FIGURE 8.4
VIIRS image from September 21, 2014, showing (a) southeastern Asia and (b) zoomed in on the Thailand coast with numerous boats just outside the Bay of Bangkok in the Gulf of Thailand. (Data courtesy of NASA, NOAA, and the Department of Defense.)

8.5.4 Subsidence

Subsidence is the gradual sinking, or caving in, of land and is often caused by human activities; for example, excessive groundwater extraction damages the foundations on which buildings are supported. Microwave data products produced using the Interferometric Synthetic Aperture Radar (InSAR) technique, described in Chapter 5, can detect millimeter-level changes in the Earth's surface over large areas and at a density far exceeding any ground-based technique. For example, the FP7 Pangeo project (https://www.pangeoproject.eu) provides free access to geohazard information, about the stability of the ground, for many of the largest cities in Europe. Figure 8.5a shows an example Pangeo product for Rome, Italy, downloaded as a shapefile and imported into QGIS for viewing. Figure 8.5b is an image of the ground deformation in Mexico City from Sentinel-1A data acquired between October 3 and December 2, 2014.

InSAR techniques are complex forms of remote sensing and will not be covered in detail in this book or through practical exercises. If you're interested in exploring this further, then the European Space Agency (ESA) provides training material, for which the hyperlink is included in Section 8.8.

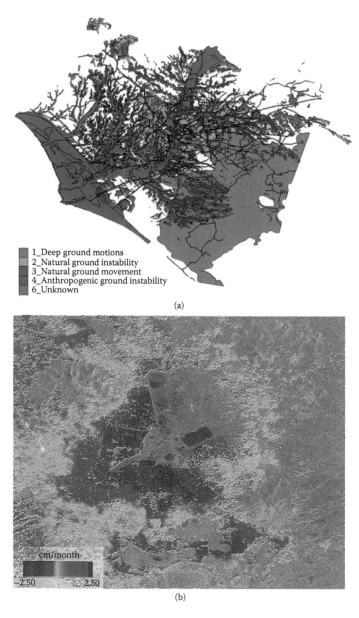

(a)

1_Deep ground motions
2_Natural ground instability
3_Natural ground movement
4_Anthropogenic ground instability
6_Unknown

(b)

cm/month

-2.50   2.50

FIGURE 8.5
(a) Example Pangeo product for Rome, Italy, and (b) Sentinel-1A data acquired between October 3 and December 2, 2014, combined to create an image of the ground deformation in Mexico City. (Sentinel-1A Copernicus data [2014]/ESA/DLR Microwave and Radar Institute—SEOM InSARap study. Pangeo product Copyright © 2012. Reproduced with the permission of the rights holders who participated in the EC FP7 PanGeo Project (262371) and the European Environment Agency. Details of the rightsholders and the terms of the license to use PanGeo project data can be found at http://www.pangeo project.eu.)

8.6 Practical Exercise: Spectral and Thermal Signatures

This practical focuses on distinguishing green spaces and buildings, together with temperature dynamics. Therefore, we'll be looking at a selection of imagery collected over New York and its wider area, often referred to as the New York Bight region, with a center of latitude 40.7127°N and longitude 74.0059°W or WRS-2 path 13, row 32, for September 8, 2002. In all the application chapter practicals, we'll be taking you step by step through the process supported by images to show the results we'd anticipate from the processing, so you can check whether you are doing everything correctly.

In this practical, we'll start with processing Landsat data for urban areas before moving on to downloading and processing MODIS data and combining it with Landsat before finally talking about ASTER data; however, this will be a discussion rather than the practical side. However, rather than using GloVis, we will use EarthExplorer to show a different way of finding and downloading data. This change of downloading platform is because the MODIS data aren't available on GloVis.

We'll also use a different Landsat data set, the L2 data rather than less processed Level 1 (L1) data. The main difference is that the L2 data have been corrected for the effects of the atmosphere through an atmospheric correction that provides greater consistency between images, unlike L1, where these inconsistencies are retained in the data; atmospheric correction will be discussed in more detail in Chapter 9. As these corrections are made, L2 is provided as surface reflectance measurements. It also includes additional data sets such as surface temperature, which is ideal for this practical's focus on temperature dynamics.

8.6.1 Step One: Downloading, Importing, and Processing Landsat Optical Data to Determine Green Spaces

We will start by downloading the Landsat data from the USGS EarthExplorer website (https://earthexplorer.usgs.gov/). If you go to the website, you'll see a browser window similar to GloVis, with various selection criteria down the left sidebar and the main view. If you click on Login, in the top right corner, you can log in to EarthExplorer using the same username and password you used to log in to GloVis.

Once logged in, the instructions to find the Landsat data are as follows:

- On the first tab, Search Criteria, click on the ∨ at the end of Feature (GNIS) and switch to the Path/Row option, and select the location using WRS-2, with the Path "13" and Row "32". Then press the Show button to move the map to this location. The map will move, and a blue pin will highlight the selected location.

- Move to the Data range at the bottom, and enter a range covering August 01–October 31, 2002. **NOTE:** This platform uses the American date format of mm/dd/yyyy to enter the dates.

- Next, move to the Data Sets tab, where you'll be presented with a wide range of potential data sets. Click on the + next to Landsat to expand the multiple options available. For this practical, we'll use the Landsat Collection 2 Level-2 data set that, if you expand, you'll see options for the different Landsat missions. We'll use Landsat-7, so check the box for this data set, and press the Results button at the bottom of the screen.

- You'll be offered six Landsat scenes for this area, together with their entity ID, acquisition date, path, and row. You could scroll through these scenes to find the image wanted, in a similar way to GloVis; alternatively, it's possible to examine the scenes to ensure which one matches your needs through the first few buttons at the bottom of each scene.

  - Show Footprint: This puts the Landsat scene location over the main map.

  - Show Browse Overlay: Overlays the actual Landsat scene over the main map.

  - Compare Browse: Select multiple scenes, and then the comparison is shown via the Show Results Control drop-down menu above the scene details. It gives options for displaying the footprints or the Landsat images.

  - Show Metadata & Browse: Brings up a box with a LandsatLook Natural Color preview image and all the image attributes.

Looking at the scenes, you'll see some have a high level of cloud, and others have partial cloud; this is because we did not set any cloud cover criteria, which can be done on the Additional Criteria tab with an option Land Cloud Cover. Be aware this is land cloud cover and does not remove images with clouds just over the sea. However, September 08, 2002 is cloud-free, and that is the one we'll use. Press the Download Options icon, a downward green arrow on top of a computer hard disk, and you'll be presented with a pop-up window with Product Options, click on the downward pointing triangle, and you'll get all the individual files to select whatever you want. As we want all the files, click on the Landsat Collection 2 Level-2 Product Bundle at the top of the window to start the download. If you only want individual files, you can select them from the options.

The downloaded file will be a 393 MB TAR zipped file with the name LE07_L2SP_013032_20020908_20200916_02_T1.tar, and hopefully, you'll recognize the different elements of the file name from the GloVis version. Unzip the tar file as expected. To import it into SNAP, there is an option to specifically import Landsat Collection 2 Level-2 data. Go to File > Import > Optical Sensors > Landsat > Landsat Collection-2 Level 2 (GeoTIFF), navigate to the data you've just unzipped, and import the MTL file.

Expand the layer, click on Bands and then export the data into a Geostationary Earth Orbit Tagged Image File Format (GeoTIFF) file using the GeoTIFF/BigTIFF option. Although it is possible to create a single GeoTIFF, it will create a large file of around 1GB. Therefore, similarly to Section 7.7.1, it is easier to subset the data to create a smaller file to work with. Click on Bands and go to File > Export> GeoTIFF/BigTIFF, and at the export dialog box, click on the Subset button over on the right side, and from the Subset dialog box, select the Band Subset tab. Similar to before, click the checkbox to "Select none", which removes all the bands, but this time add back B1, B2, B3, B4, and B5, and click Okay. Change the location of the saved file if required, and click "Export Product". This process should produce a GeoTIFF file of approximately 289 MB. Finally, import this new GeoTIFF file into QGIS by adding it as a raster layer through the menu item Layer > Add Layer > Add Raster Layer, following the process outlined in Section 7.5, which gives a black-and-white view.

The next action is to create a pseudo-true-color composite, and the best bands from Landsat-7 ETM+ data to do this are in the order of B3, B2, and B1 (https://www.usgs.gov/media/images/common-landsat-band-rgb-composites). Go to the Symbology tab in Properties, and put the order of B3, B2, and B1, and in addition, set the "Additional no data value" to zero in the Transparency tab. You should end up with an image similar to Figure 8.6a.

However, in this chapter, we're interested in temperature, and using a combination of visible and NIR wavebands, we can create a false-color composite using the band 4, 3, and 2 combination. This type of false-color image allows one to separate natural and artificial features, even though Landsat-7 ETM+ data were not designed for urban applications. If you change the layers in the Symbology tab to B4, B3, and B2. Then move to the Histogram tab, and compute the Histogram. Stretch all three bands to the Min/Max of 7,000–11,000. The detailed instructions can be found in Sections 7.5 and 7.6. This stretch should produce an image like the one shown in Figure 8.6b, where the NIR/red/green composite separates the land from the dark water. If you zoom into the New York Bay area, you'll see many single bright pixels representing boats coming in and out of the area. In addition, if you zoom in on the Hudson River, you can see a sediment plume visible going up the river that is green in color with swirls.

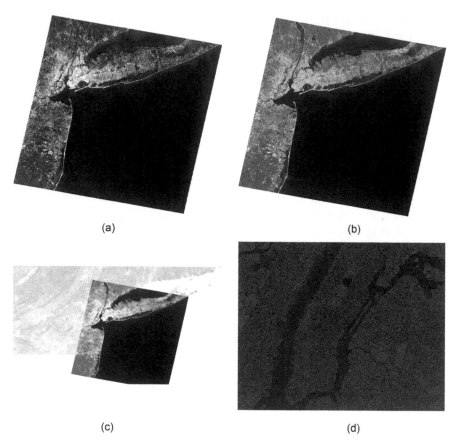

FIGURE 8.6
Collection of imagery collected over the New York Bight on September 8, 2002, as (a) the Landsat-7 ETM+ pseudo-true-color composite, (b) the Landsat-7 ETM+ false-color composite, (c) the Landsat-7 ETM+ false-color composite with the MODIS daytime land surface temperature overlaid after import, and (d) zoomed-in ASTER false-color composite for New York and Manhattan. (Data courtesy of NASA/USGS.)

8.6.2 Step Two: Downloading and Importing MODIS Data to QGIS

The second data set we're going to use is from MODIS, which is a twin satellite mission, with MODIS-Terra crossing the equator in the morning and MODIS-Aqua crossing it in the afternoon. On EarthExplorer, go to the same New York location as in the previous part of the practical. However, rather than using Landsat, select the Data Sets tab. Uncheck the Landsat-7 box to remove it, and close down the Landsat options by clicking the - symbol. This time, expand the NASA LPDAAC Collections, and from the drop-down list, select the MODIS Land Surface Temp (LST) and Emiss – V6.1 > MODIS MOD11A1 V6.1.

At this point, you'll be told that you must sign in with your NASA Earthdata Login Credentials and USGS EarthExplorer credentials. These are free to create, and to register, you'll need to create a username and password, and enter some basic contact and usage information. Once you have registered and logged in, press the Results button at the bottom of the left sidebar, the details of the available data sets will appear. You may have a lot of available data sets if you kept the time period the same as for Landsat, as there should be one data set available daily. To ensure consistency with the Landsat data, find the data set for September 8, 2002.

Click on the Download button, and you'll get a pop-up offering two Download options: the HDF Format and Metadata. It's only necessary to download the HDF file, which is Hierarchical Data Format, and it will download the file MOD11A1.A2002251.h12v04.061.2020129135939.hdf.

As this is not a zip file, it can be loaded straight into QGIS using the menu item Layer > Add Layer > Add Raster Layer. After you click Add, you'll be shown a list of data sets in the file – you may find it easier to expand the size of the "Select Items to Add" box by dragging its edge to see all of the data sets available. You can select all the data sets, individual ones, or multiple data sets by holding down the CTRL key and selecting more than one. For this part of the practical, first press the Deselect All button, and then using the CTRL key select the following:

- MODIS_Grid_Daily_1km_LST:LST_Day_1Km, which is the daytime LST and
- one of the emissivity products either
  - MODIS_Grid_Daily_1km_LST:Emis_31 or
  - MODIS_Grid_Daily_1km_LST:Emis_32

Press Add Layers to import these data into QGIS.

NOTE: The Land Processes Distributed Active Archive Center (https://lpdaac.usgs.gov/product_search/?collections=Combined+MODIS&collections=Terra+MODIS&collections=Aqua+MODIS&status=Operational&view=list) lists all the available MODIS terrestrially focused products alongside their file type and spatial and temporal resolution.

8.6.3 Step Three: Combining MODIS Thermal Data with Optical Data from Landsat

Rearrange the layers by dragging them, so that the MODIS_Grid_Daily_1km_LST:LST_Day_1Km layer is on top, followed by the Landsat layer, and then the MODIS Emissivity layer is at the bottom – you can expand the size of the left sidebar by dragging its edge to see the full data set names. The main visible layer will be MODIS_Grid_Daily_1km_LST:LST_Day_1Km layer as it's on top; go to the Properties of this layer and, under the Transparency tab, use the slider or the number box, to set the "Global Opacity" to 50%; click OK. Landsat will now become visible through the MODIS layer. Go to the Histogram tab on the properties for the MODIS layer, and Compute the Histogram. Change the Min/Max for the stretch to 284 and 310; these numbers refer to LST in kelvin (K), rather than the surface reflectance levels of the Landsat image. Converting the temperature to Fahrenheit indicates that the LST ranges between 55.5 and 98.3°F. Finally, switch the emissivity layer off by unchecking the box next to that layer name.

The image you now see will be similar to Figure 8.6c, where the MODIS image shows the LST with brighter areas indicating warmer temperatures, which are approximately equivalent to the non-vegetated areas on the Landsat image and, therefore, likely to correspond to buildings/impervious surfaces, with the darker areas representing the cooler vegetation. This is easier to compare by using the X to turn the Landsat layer on and off. If you zoom in, the MODIS layer becomes fuzzy compared with the Landsat layer because of the difference in spatial resolution: MODIS has 1 km pixels, and Landsat has 30 m pixels.

Using the checkboxes to turn layers on and off will also allow you to compare the emissivity image with the MODIS LST or the Landsat false-color composite. Emissivity is effectively the inverse of the MODIS LST image, with darker areas representing the urban features and lighter areas for the more vegetated areas. So the symmetry between the images will be apparent. You will be able to see the same urban versus vegetated areas highlighted on the Landsat image.

8.6.4 Step Four: Comparing Thermal Data from Landsat and MODIS

An alternative comparison is to display the Landsat-7 thermal waveband six and compare it with the MODIS LST data.

Go to the menu item Layer > Add Layer > Add Raster Layer, navigate to the individual GeoTIFF files downloaded at the start of this practical, and import the single waveband from the file "LE07_L2SP_013032_20020908_20200916_02_T1_ST_B6".

Go to the Symbology tab, use Min/Max values of 42,000 and 43,000 for the contrast stretch, and set the "Additional no data value" to zero. Order the layers so that the MODIS LST is on top, followed by the single Landsat thermal waveband, and turn off the remaining layers.

We're now comparing the MODIS_LST image in kelvin, with the Digital Number (DN) of the Landsat thermal band. The Landsat L2 products are stored as integer values, meaning the files are smaller than they would be if float values were stored. To convert from DN to kelvin you need to multiply by 0.00341802 and then add 149.0; with these values being found in the MTL file, you can undertake using the Raster > Raster Calculator using the following equation:

if ("LE07_L2SP_013032_20020908_20200916_02_T1_ST_B6@1" > 0, ("LE07_L2SP_013032_20020908_20200916_02_T1_ST_B6@1" * 0.00341802) + 149.0, 0)

Once you have entered the calculation in the Raster Calculation Express Field, press the ... at the end of Output Layer, and give the GeoTIFF a name such as "Calculation" alongside a location for the file to be saved. Once you've done this, the OK button will no longer be grayed, and you can click this OK button. This process will create a new layer called "Calculation" which is now in the Landsat LST in kelvin, and you will see that the min/max figures are similar to the MODIS values.

In both images, the brighter areas indicate hotter temperatures, and the darker areas indicate cooler temperatures. This means that the images look very similar, and you could use either to look at the temperature variability within an area.

8.6.5 Step Five: Example of ASTER Data

We'll end the practical with a description of using ASTER, which is also onboard Terra, for both urban land cover classification and agricultural studies where areas of interest may be too small to be well resolved by sensors such as MODIS.

The ASTER library of spectral signatures, discussed in Section 2.2, was developed by scientists working on data on ASTER. This sensor contains a five-waveband TIR (8–12 μm) scanner with a 60 km field of view at 90 m spatial resolution, and three visible/NIR wavebands at 15 m spatial resolution (Abrams et al., 2015). Therefore, as a comparison, a visible/NIR color composite of New York on the same date as the MODIS data are shown in Figure 8.6d, with the image zoomed in so that you can see the details of New York and Manhattan with Central Park being the red-colored rectangle in the middle; it uses wavebands 3, 2, and 1 to show vegetation in red and the urban regions in blue/gray.

Although the ASTER data set is available to download via EarthExplorer, it requires additional processing to allow the georeferencing to be correctly interpreted by QGIS. Therefore, we've only given the image as an example, and if you're interested in taking this more complicated next step, further details on how to do this are available on the associated learning resource website (https://playingwithrsdata.com/).

8.7 Summary

This is the first of the practical chapters looking at remote sensing within urban environments. The theory part of the chapter delved more deeply into the optical and thermal signatures of different materials on Earth and how they can be used to interpret satellite data.

We've introduced you to five real-world application areas, identifying how these work and what data to use. We finished with the first detailed practical exercise, using MODIS and Landsat data, to explore an example urban environment's spectral and thermal signatures. You now have the skills to undertake a remote sensing analysis of your town and city; you never know what you might discover!

8.8 Online Resources

- Associated learning resource website: https://www. playingwithrsdata.com/
- FP7 Pangeo project: https://www.pangeoproject.eu/
- Land Processes Distributed Active Archive Center (LP DAAC): https://lpdaac.usgs.gov/product_search/?collections=Combined +MODIS&collections=Terra+MODIS&collections=Aqua+MODIS &status=Operational&view=list
- Millennium Ecosystem assessment: https://www.millenniumas-sessment.org/
- USGS Common Landsat Band RGB Composites: https://www. usgs.gov/media/images/common-landsat-band-rgb-composites/
- USGS EarthExplorer website: https://earthexplorer.usgs.gov/

8.9 Key Terms

- Blackbody: A hypothetical perfect radiator of energy.
- Brightness temperature: Temperature a blackbody in thermal equilibrium with its surroundings would have to be to duplicate the observed intensity of the feature of interest.
- Ground radiance: The solar irradiance reflected after striking the ground.
- Irradiance: EM radiation that's dispersed across an area.
- Radiance: A beam of EM radiation reflected or scattered in a particular direction.
- Sensor radiance: The reflected radiance that reaches the satellite sensor.

- Solar irradiance: EM radiation emitted by the sun toward the Earth.
- Solar zenith angle: Vertical angle of the incoming solar irradiance.
- Urban environment: A town or city, and the natural and manmade features that surround it to support the human population.

References

Abrams, M., H. Tsu, G. Hulley et al. 2015. The Advanced Spaceborne Thermal Emission and Reflection Radiometer (ASTER) after fifteen years: Review of global products. *Int J Appl Earth Obs Geoinf* 38:292–301.

Baldridge, A. M., S. J. Hook, C. I. Grove and G. Rivera. 2009. The ASTER spectral library version 2.0. *Remote Sens Environ* 113:711–715.

Coll, C., J. M. Galve, J. M. Sánchez and V. Caselles. 2010. Validation of Landsat-7/ETM+ thermal-band calibration and atmospheric correction with ground-based measurements. *IEEE Trans Geosci Remote Sens* 48(1):547–555.

Daily, G. C. 2013. "Nature's services: Societal dependence on natural ecosystems (1997)". In *The Future of Nature: Documents of Global Change*, eds. L. Robin, S. Sörlin and P. Warde, 454–464. New Haven: Yale University Press.

Elvidge, C. D., P. Cinzano, D. R. Pettit et al. 2007. The Nightsat mission concept. *Int J Remote Sens* 28(12):2645–2670.

Hofmann, P., J. Strobl and A. Nazarkulova. 2011. Mapping green spaces in Bishkek—How reliable can spatial analysis be? *Remote Sens* 3:1088–1103.

McGranahan, G. and P. Marcotuullio. 2005. Urban systems. In *Ecosystems and Human Well-Being: Current State and Trends*, Vol. 1, eds. R. Hassan, R. Scholes and N. Ash, 797–825. Washington, DC: Island Press.

Wittich, K.-P. 1997. Some simple relationships between land-surface emissivity, greenness and the plant cover fraction for use in satellite remote sensing. *Int J Biometeorol* 41(2):58–64.

9

Landscape Evolution

As discussed in Chapter 8, changes in urban environment ecosystems often happen in small areas within a city and over relatively short periods. Changes in the rural environment are different and, if anything, the symbiotic relationship between humans and the natural environment is stronger in the rural ecosystem. Changes in this landscape may occur over much larger areas, possibly thousands of kilometers, and over a much longer time span. Remote sensing is a valuable monitoring tool, as it can use its multiple temporal and spatial scales to develop landscape indicators, such as vegetation indices, to help assess these changes.

There are many classification systems for describing the Earth's surface, with Land Use (LU) and Land Cover (LC) being two dominant surface descriptors. As they're strongly interconnected, most classification systems mix them and use the term *Land Use and Land Cover* (LULC). However, it's important to note that Earth Observation (EO) data alone will not be able to distinguish all the different flora and fauna species and populations; thus, ideally, it needs to be combined with other data sets and knowledge.

9.1 Principles of Using Time-Series Analysis for Monitoring Landscape Evolution

To monitor LULC change, it's necessary to see how the land evolves over time. To gain this understanding requires a combination of historical archives and new satellite data acquisitions. There are several potential issues to be aware of when comparing data over time:

- Different sensors may be used on different missions, such as Landsat Multispectral Scanner versus Landsat Thematic Mapper (TM), meaning that the waveband specifications and radiometric sensitivity may not be identical. So the images may not be directly comparable.
- Combining different satellites, such as Landsat and Sentinel, can cause issues as spatial resolution, waveband specifications, and

radiometric sensitivity will not be the same, making it difficult to directly compare them. Therefore, there are international activities to harmonize these datasets and so make this process easier, for example, the US Harmonized Landsat Sentinel-2 (https://www.earthdata.nasa.gov/esds/harmonized-landsat-sentinel-2) and European Sen2Like (https://github.com/senbox-org/sen-2like) project, which is a tool to generate Sentinel-2 and Landsat harmonized products.

- The environmental conditions at the time of acquisition can substantially impact the resulting data. For example, Thomson (1992) showed that two Landsat TM images collected 16 days apart resulted in a change of 6° in the solar zenith angle (owing to the change in date), preventing accurate comparisons for steeply sloping areas. Therefore, corrections for the atmosphere are ideally required. However, when a correction for the atmosphere, termed atmospheric correction, has been applied, the product is Level 2, rather than Level 1 (L1), and such products can be downloaded for multiple sensors, including Landsat and Sentinel-2. We'll also use a Plugin in Quantum Geographic Information System (QGIS) in this chapter's practical to apply an atmospheric correction.

- The geometrical (positional) corrections of pixels need to be sufficiently accurate so that errors when comparing different images are minimized; for example, if the geometric errors are large, then the same position on the Earth may not be represented by corresponding pixels in different images. It's assessed by using the root-mean-squared error (RMSE), which is a statistical value representing the overall positional error for an image; ideally, it should be smaller than half a pixel. If images don't match well, an image-to-image registration can be applied by selecting the same features on the processed image and reference image/map and applying a mathematical transformation to force one to match the other. For Landsat scenes, the MTL file contains the RMSE values for the x and y directions as well as a combined value. For the Sentinel-2 data, this information is in the GEOMETRIC_QUALITY.xml file in the DATASTRIP/*Name*/QI_DATA folder. Also, geometric condition improvements are being made with the introduction of geometry files within the L1 Landsat data, and similar data are present in Sentinel-2 products that allow a geometric correction to be calculated; although it is not so commonly applied. However, it is another area of focus for projects such as HLS and Sen2Like. These developments should mean that soon geometric issues won't be things data users need to be concerned about.

- Imagery should also be used with the same projection and datum, and if the area of interest has dramatic variations in terrain, an orthorectification should also be considered; Landsat scenes have this applied if they've undergone the full geometric processing, as described in Section 6.5, and for Sentinel-2 this is applied when data is acquired over the land.
- Dates also need to be carefully selected such that phenological conditions (seasonal patterns) are not significant or are corrected by using a mathematical model.

For these reasons, if a time-series analysis is being performed, it's better to use data processed in a delayed versus near-real-time mode, if possible, as the delayed data production will allow for more accurate calibration in terms of both the geometric and radiometric accuracy. These issues also mean that data archives are often reprocessed as new calibration knowledge is gained to make the data as accurate as possible. Hence, data downloaded around 2 years ago may not be the same data available to be downloaded today. As you'll have seen earlier in the book, Landsat underwent a large data reprocessing in 2020. The Landsat archive is being continually updated and improved, and you can find the latest developments on the Landsat news page (https://www.usgs.gov/landsat-missions/news).

One of the advantages of the Landsat archive, and the Sentinel-2 Level 1C products, is that all these corrections have been applied, and the data are supplied in a consistent projection. However, it's still important to carefully check the type of L1 data used because, as discussed in Section 6.4, different scenes can have different levels of geometric correction.

9.2 Landscape Evolution Techniques

Once you're satisfied images can be compared, then there's a choice of techniques that can be applied to help monitor landscape changes:

The simplest is to create a difference image by applying a mathematical algorithm to subtract one image from another or calculating the ratio of two images or even just visually flicking back and forth between them. These approaches are easy to implement but susceptible to radiometric and geometric errors.

- Assuming geometric differences can be reduced, the change detection can be improved by using a vegetation index rather than radiometric values. These indices enhance vegetation differences, such as green vegetation's differential absorption and reflectance of solar irradiance (Deering and Haas, 1980), while reducing the impact of atmospheric differences. An example of this type of index, the Normalized Difference Vegetation Index (NDVI) was part of the practical in Chapter 6 and will be reviewed in more detail in Section 9.3.

- Unsupervised and supervised land classification techniques are where pixels are grouped based on common characteristics, such as the spectral shape, and then assigned a land classification. In the unsupervised version, a computer algorithm performs the classification. Then the user attempts to interpret the results. In contrast, in the supervised technique, the user defines a library of classifications, known as training sites, which the computer then uses to assign pixels according to the best statistical matches.

- Principal component analysis (PCA) is a specific type of unsupervised classification that reduces large data sets to their key variables using vector analysis to determine the pixels with the most significant variability. The process creates a new output where those pixels, or groups of pixels, with the greatest variance are known as the first component (PC1), those with the second greatest variance are the second component (PC2), and so on. This methodology reduces large data sets, with many variables, down to the few key variables that can be used as a proxy to interpret the whole data set. This approach is beneficial when using hyperspectral data as there's a large number of wavebands or time-series data sets that have strong correlations within them. PCA was first applied to LU change by Byrne et al. (1980), who showed that the statistically minor changes (i.e., PC3 and PC4) were associated with local changes in land cover, which was also seen by Richards (1984), who showed it's important not to simply focus on the highest components.

9.3 Optical Vegetation Indices for Landscape Evolution

The most common use of optical data within land mapping surrounds the construction of indices, which focus on using "greenness" or vegetation, including using the classification techniques described in Section 9.2. There's also a move toward first segmenting the image so that it's composed of homogeneous multipixel objects, rather than individual pixels, which ideally correspond to real-world features; segmentation works by using not only the spectral shape but also properties such as texture and size.

As mentioned in Section 9.2, vegetation indices give quantitative measurements of photosynthetically active vegetation by exploiting specific spectral reflectance characteristics, often depending on a high reflectance in the near-infrared (NIR) by plant matter contrasting to the strong absorption by chlorophyll-a (Chlor-a) in the red wavelengths that's termed the *red edge* (Curran 1989). NDVI is a well-known example, and Figure 9.1 shows a SPOT-VGT NDVI composite for the Africa continental tile derived at 10-day temporal resolution, downloaded from the Copernicus Land Monitoring Service (https://land.copernicus.eu/). There are many other indices, such as the Enhanced Vegetation Index or the Normalized Difference Water Index developed by Gao (1996), that are used to characterize the amount of water held in vegetation from the NIR divided by the mid-infrared (MIR) waveband ratio:

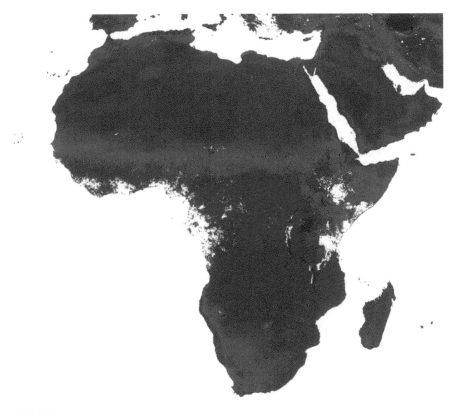

FIGURE 9.1

May 13, 2014, 10-day composite of NDVI derived from SPOT-VGT data for the Africa continental tile. (Copyright Copernicus Global Land Service, 2013. Distributed and produced by VITO NV, Belgium.)

$$NDWI = \frac{(NIR - MIR)}{(NIR + MIR)} \tag{9.1}$$

Firstly, this index doesn't completely remove the background soil reflectance effects, similar to NDVI. However, the information about vegetation canopies in the 1.24-μm waveband is very different from that in the red waveband, which gives additional information and is an interesting independent vegetation index.

Secondly, as Medium Resolution Imaging Spectrometer (MERIS) has narrower wavebands than sensors such as Landsat and VGT, it can also produce indices such as the MERIS Terrestrial Chlorophyll Index (MTCI) that uses wavebands situated along the vegetation "red edge"; for example, waveband 9 is at 708.75 nm; global products are available from the UK-based Centre for Environmental Data Archival (CEDA), alongside many other useful datasets. (https://catalogue.ceda.ac.uk/uuid/9ed8d70 ffde5b757691f3d8124f13148).

Therefore, as different indices use different spectral bands, not all indices can be applied to all sensors. The Index DataBase developed by Henrich et al. (2012) gives an overview of 66 different vegetation indices, including which indices match with which sensor. These differences mean that the performance of indices is always different, and so it's important to consider the theoretical background, validity range, and purpose when creating results that are intended to be compared spatially and temporally, as noted by Clevers (2014).

Despite vegetation indices being simplistic approaches, they are often highly correlated with biophysical variables such as the leaf area index (LAI), fraction of Absorbed Photosynthetically Active Radiation (fAPAR), and vegetation cover fraction (fCover), which are all classed within the terrestrial group of the essential climate variables (ECVs) used by the United Nations (UN) to monitor climate change. Therefore, these sorts of indices can be particularly useful environmental assessment tools.

9.4 Microwave Data for Landscape Evolution

The 24-hour, all-weather capability of microwave data, such as Synthetic Aperture Radar (SAR), makes it ideal for monitoring changes in regions with high cloud cover such as tropical forests as it can penetrate through both precipitation and smoke. Tropical forests cover huge areas and are under pressure from industrial and socioeconomic factors, such as land scarcity and agricultural intensification, with the effects including deforestation and degradation. Therefore, SAR is a valuable source of information for LULC change alongside ground surveys, aerial photography, Lidar, and optical satellite imagery.

Like wavebands in optical imagery, features are seen differently by different microwave bands or frequencies. For example, when a tree is observed, the X-band radiation is backscattered from the leaves; C-band, from the small branches; L-band, from the small to large branches; and the P-band, from the large branches and the trunk (Wang et al., 1995). Hence, shorter frequencies penetrate further through the tree canopy and are more likely to interact with the ground. Also, water and roads will tend to appear dark while vegetation is brighter because of their high moisture content, higher dielectric constant, and a rougher surface creating greater backscatter.

In addition to the frequency band, polarization also has an impact; as described in Section 3.2.4. Single-polarized vertically (VV) energy does not pass as readily through forests as single-polarized horizontally (HH) energy because vertically oriented structures such as the tree trunks interact more with vertically polarized energy and reduce penetration. In contrast, cross-polarized data (HV or VH) is sensitive to depolarization, where the horizontally polarized energy is transformed into vertically polarized energy upon striking an observed object, or vice versa; this occurs mainly over vegetation, but hardly ever over open ground. Therefore, by creating a multifrequency color composite (HH, VV, and HV as red, green, and blue, respectively), different surfaces will become separated.

9.5 Landscape Evolution Applications

9.5.1 Mapping Land Cover

Humans have used land in various ways during their existence, but the industrial and technological revolutions of the last century have significantly increased the pressure on land usage and the speed at which large areas can have their use changed. This has altered the planet's land cover with implications of gradual desertification in certain areas, the disappearance of forests, poor farmland use, and gradual drying up of wetlands. To monitor and manage the natural environment, overall maps of LULC have become increasingly important, and remote sensing offers a route for developing and maintaining such maps.

Europe is leading the way in mapping this type of land cover with the CORINE Land Cover project, implemented by the European Commission between 1985 and 1990. It provided a set of methodologies and nomenclatures (systems of names) for classifying 44 different types of land

(Buttner, 2014), enabling Europe-wide maps to be produced. An example of a CORINE land cover map can be seen in Figure 9.2a based on 1990 data at 250 m resolution showing central Spain, and Figure 9.2b is from 2006, showing a zoomed-in map around Madrid.

Vast parts of the map are in green showing various forests and vegetation, with the

- Bottom left corner has brown polygons representing vineyards, fruit trees, and olive groves
- Top left corner is Madrid with its urban land classification
- Buendia reservoir in blue in the center of the image
- Small areas of black pixels represent burnt areas, although these are not represented on the legend

These maps are derived from satellite data, and a series of CORINE Land Cover data sets (1990, 2000, 2006, 2012, and 2018) are available through the Copernicus Land Service (https://land.copernicus.eu/pan-european/corine-land-cover) or directly from the European Environment Agency. The 1990 and 2000 maps were based on Landsat data, the 2006 map was based on SPOT-4/5 and Indian Space Research Organisation's ResourceSat1, and the 2012 data set was based on ResurceSat1 and RapidEye. Repeat coverage is essential to remove the effects of vegetation phenology and identify land cover changes; with the following and future updates also using the Sentinel-2 satellites, it should benefit from its more frequent revisit time. We'll use CORINE dataset as part of the practical in Chapter 12.

In 2015, Rwanda released its National Land Use Planning Portal, the first in Africa, with the primary objectives of disseminating national and district LU plans to the public, facilitating access to information related to LU planning, and increasing awareness and education on LU planning. A more recent focus has been Analysis Ready Data (ARD), which is satellite-derived data that should be easily useable for applying algorithms and extracting information so it can be readily used for multiple applications, including LU planning. Digital Earth Africa (https://www.digitalearthafrica.org/) aims to provide a routine, reliable and operational service, using EO to deliver decision-ready products enabling policymakers, scientists, the private sector, and civil society to address social, environmental, and economic changes in the continent and develop an ecosystem for innovation across sectors.

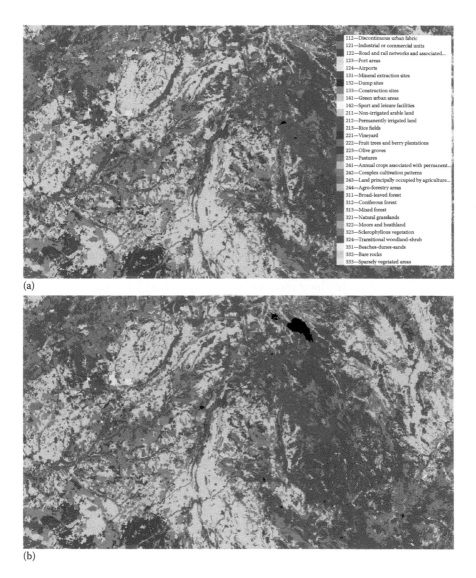

FIGURE 9.2

A comparison of the (a) 1990 and (b) 2006 CORINE Land Cover 250-m resolution raster data zoomed in to show central Spain with changes, including the urban growth of Madrid (red/pink pixels) and burnt areas (black pixels). (Courtesy of the European Environment Agency.)

9.5.2 Agriculture

Mapping of agricultural crops using aerial photography has been occurring since the 1930s, with pioneering work on remote sensing in agriculture during the 1950s and 1960s (Duane Nellis et al. 2009). The need for agriculture to be sustainable and not to damage the natural environment is increasingly important. Also, knowing the amount and layout of non-cropped areas allows for the estimation of the background level of biodiversity, which can help determine if the damage is being caused to local ecosystems. A financial application of this mapping type is used in the European Union, where at least 5% of farmers claim an annual subsidy through the Common Agricultural Policy. These claims are checked primarily using very-high-resolution optical imagery because of the increasing need to check small landscape elements such as hedges or isolated trees (Richardson et al., 2014). The European Space Agency (ESA) funded Sentinels for Common Agriculture Policy (Sen4CAP) project (http://esa-sen4cap.org/) has demonstrated how the Sentinels could take the policy into a new era through operational and systematic observations that identify the type of crop planted, how it grows through the season and when it is harvested.

In China, rapid urbanization has meant there's less land available to grow food; growing grains, including rice, wheat, and corn, are increasingly carried out on low-quality land more vulnerable to crop failure. Recent research by Chen and Mcnairn (2006) using Radar imagery identified that backscatter increases significantly during the short periods of vegetation growth for rice fields and allows them to be separated and mapped from other land cover types. Mapping the geographic distributions of crops and characterizing cropping practices have tended to involve waveband ratios, together with supervised and unsupervised classification techniques using medium- to very-high-resolution optical imagery depending on the area to be covered and the level of detail needed. The availability of these data is increasing as commercial companies launch constellations alongside space agency-provided missions such as Landsat and Sentinel. For example, Chang Guang Satellite Technology's Jilin-1 constellation, Maxar's QuickBird, WorldView, and GeoEye-1 satellites, Planet Lab's constellation of Dove and Pelican satellites, and Satellogic's Aleph-1 constellation. There is also a focus on fusing optical and radar imagery to deal with missing data due to clouds.

This chapter's practical exercise in Section 9.6 provides an approach, using QGIS, for creating a land classification using Landsat.

9.5.3 Forestry and Carbon Storage

Monitoring forest biomass is essential for understanding the global carbon cycle as forests account for approximately 45% of terrestrial carbon and sequester significant amounts of carbon yearly. Therefore, there's a strong interest in monitoring and estimating carbon stocks in forests, driven by the potential to mitigate greenhouse gas emissions through programs such as the UN Intergovernmental Reducing Emissions from Deforestation and Forest Degradation in Developing Countries (REDD+) program. For example, oil palm is a very important resource for Indonesia and Malaysia; however, the expansion of oil palm plantations has led to the degradation of the peat swamp forest ecosystems as well as primary and secondary forests, which has contributed to Indonesia being classed as the third largest greenhouse gas emitter by the World Bank.

The REDD+ program was secured in 2013 during the 19th Conference of the Parties to the UN Framework Convention on Climate Change and required countries to map and monitor deforestation and forest degradation, plus develop a sustainable forest management system.

Remote sensing can support these activities with Phased Array type L-band Synthetic Aperture Radar-2 (PALSAR-2) data from the ALOS-2 satellite, optimized for biomass estimation by having data at different polarizations, incidence angles, and spatial resolutions. In addition, multiple-date dual polarization data (HH and HV) can allow for textural rather than just signal magnitude processing as Radar reflections arise from the top of the canopy and changes (e.g., missing trees owing to deforestation) alter the return signal by creating shadows. A ground resolution of around 5 m is needed to see small-scale changes, such as selective logging, with coarser resolutions showing larger areas of degradation.

Figure 9.3 shows an area crossing the Malaysia/Brunei border, captured on September 10, 2014, using ScanSAR mode, which has a swath width of 350 or 490 km, at a spatial resolution of 100 or 60 m, respectively. Figure 9.3a is HH polarization, while Figure 9.3b is HV polarization (this sample product is described as "Sarawak, Malaysia" on the ALOS-2 website). Malaysia is on the left of the images, with Brunei on the right and the border to the river's right. It's noticeable that Brunei appears to have an untouched forest compared with the more disturbed forest in Malaysia, which shows up starkly as

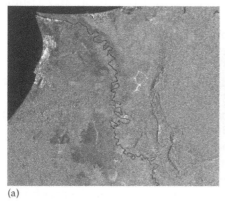 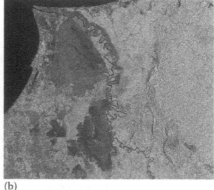

(a)　　　　　　　　　　　　　　(b)

FIGURE 9.3

An area crossing the Malaysia/Brunei border, captured on September 10, 2014, by PALASAR-2 as the (a) HH and (b) HV polarizations. (Original data are provided by JAXA as the ALOS-2 sample product, © JAXA.)

darker regions with the HH polarization in Figure 9.3b. In contrast, the HV polarization better highlights the urban areas around the coast.

An alternative technique is to analyze the infrared (IR) wavebands (e.g., Luca et al., 2002), which reveal decreases in reflectance between primary forest and sites with younger growth, which can be linked to structural development within regenerating canopies associated with the dominant pioneer species. Also, variables such as LAI, canopy cover, and direction gap fraction are used to characterize canopy structure but require accurate on-ground measurements and are extremely time-consuming to collect. Therefore, these parameters tend to be assessed indirectly or increasing using ground-based Lidar data, called terrestrial laser scanning, which includes high-end mobile phones.

9.5.4 Fire Detection

Biomass burning creates high concentrations of carbonaceous trace gases and aerosols, including CO_2 emissions, in the atmosphere, and is linked to land use changes in tropical forests, which can dominate countries' emissions compared with industrial sources. The traditional remote sensing approaches for detecting fires are based on mapping the burned areas; an example of this can be seen by returning to the CORINE Land Cover maps and in Figure 9.2b, which shows burn scars as black polygons.

More recent approaches have focused on using thermal IR (TIR) wavebands to measure fire radiative power (FRP), which is an estimation of the radiant heat output of a fire, and can be derived from both polar-orbiting and geostationary satellites. Moderate Resolution Imaging Spectroradiometer (MODIS) provides fire products such as MYD14 (thermal anomalies/fire products derived primarily from the 4- and 11-µm radiances), and as the TIR data can be acquired during both the day and night, both Aqua and Terra collect data twice a day.

Figure 9.4 shows the fires on the island of Tasmania, Australia, on January 4, 2013, using MODIS-Aqua. The background image is a top-of-atmosphere (TOA) reflectance pseudo-true-color composite using wavebands 1 (centered at 645 nm), 4 (555 nm), and 3 (470 nm) as red, green, and blue, respectively; a plume of smoke is visible down the bottom right corner of the image, although it is difficult to separate it from the clouds visually. The inset is the MYD14 product, which shows an FRP signal as the small red area in the island's southeast corner.

Therefore, to get a complete overview of a situation for fire detection, it's necessary to use a combination of optical imagery for smoke and burnt areas together with fire-related products.

Fire products are also available from the Copernicus Atmosphere Monitoring Service (https://atmosphere.copernicus.eu/) and National Aeronautics and Space Administration (NASA) Open Data Portal (https://data.nasa.gov/). The Copernicus Land Monitoring Service also offers a global burnt area product,

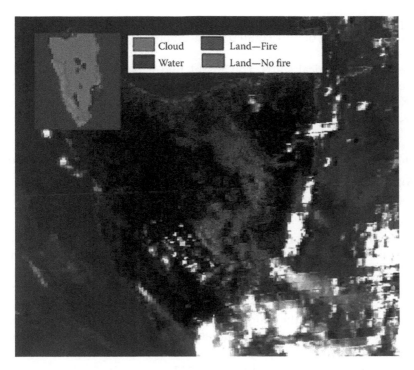

FIGURE 9.4
Fires on the island of Tasmania, Australia, on January 4, 2013, captured using MODIS-Aqua. Displayed as a pseudo-true-color composite and inset with the MYD14 product. (Data courtesy of NASA.)

which is another ECV, at 300 m spatial resolution created from Sentinel-3 data (see https://land.copernicus.eu/global/themes/Vegetation).

9.6 Practical Exercise: Supervised Land Cover Classification

The practical will undertake a supervised land classification in the US State of Kansas, near the city of Ulysses, using the "Semi-Automatic Classification" QGIS plugin. This is a complicated and challenging remote sensing practical that requires identifying and classifying features in a Landsat scene to create training sites that form the basis of the full scene land classification.

As this practical has several new processes and elements, we're going to break it down into three stages, each of which will have several steps. The first stage processes the data ready for land classification, the second stage performs a complete land classification using provided training sites, and the third stage shows you how to create your own training sites. Although it's possible to miss out the second stage, completing it will provide a better understanding of the overall process before you try setting up training sites.

Before the practical starts, you might find it helpful to watch an ESA video entitled "Planted patchwork" (https://www.youtube.com/watch?v=Rx6rwCkV0R4), which explains the agricultural features within the area of Kansas around Ulysses.

9.6.1 First Stage: Creating the Data Set Ready for Land Classification

9.6.1.1 Step One: Installing Semi-Automatic Classification Plugin into QGIS

Download the free "Semi-Automatic Classification" plugin (SCP), for which you'll need an Internet connection, by going to QGIS menu item Plugins > Manage and Install Plugins... Into the search box at the top, type "Semi" and in the list of plugins that appear you should see "Semi-Automatic Classification Plugin". If you select that link, a description of the SCP will be on the right side of the panel. Click the Install plugin button. You will see the plugin is downloaded and installed, and once completed, click Close. After installation, you'll have the "SCP & Dock" panel, beneath your layer panel, a new menu item SCP, and a new toolbar underneath the others at the top of the page.

You'll also be offered a tour of the SCP plugin, which you are welcome to do or plunge into the rest of the exercise as we'll take you through step by step. There is a web page offering help, support, and news about the SCP plugin (https://fromgistors.blogspot.com/)

9.6.1.2 Step Two: Importing and Preprocessing the Data

For this practical, we will use Landsat-8 data from June 14, 2021, for the city of Ulysses at 37.5°N 101.5°W, which equates to WRS-2 path 31, row 34. Use GloVis and the various options on the Interface Controls to find the correct dataset, and as you have done previously, download the Level-1 Product Bundle that will give you a tar file of approximately 1.2GB, with the name LC08_L1TP_031034_20210614_20210622_02_T1.tar. Once downloaded, unzip the file ready to import the data.

Rather than going through SNAP to create a single GeoTIFF, as we did in Chapter 9, we will import the data into QGIS using SCP directly.

Go to the menu item SCP > Pre-Processing > Landsat, which will open the SCP dialog box. There are several things needed to complete the preprocessing step:

- For the first box, "Directory containing Landsat bands", click the folder icon to navigate to the folder containing the extracted GeoTIFF files. Select the folder and press the Select Folder button. **Note:** as discussed in a previous chapter, using remote sensing data requires good organizational file management skills, and having a separate directory was suggested for each download; this is one of the reasons why!

- Second box, "Select MTL file", click on the file icon which will open up the folder selected in the step above "Directory containing Landsat bands", and select the MTL file in the folder – you may need to increase the size of the window by dragging the corner to be able to see the full file names. Press Open. This step will populate some of the lower fields in the dialog box.

- DOS is an acronym for Dark Object Subtraction and is a simple technique for applying an atmospheric correction that allows for the creation of bottom-of-atmosphere (BOA), rather than TOA reflectance data. Turn on the checkbox marked "Apply DOS1 atmospheric correction" by clicking on it, and a tick will appear in the checkbox. If you can't see the full name for this checkbox and can only see "Apply DOS1 atmos" this is the right checkbox; if the overall dialog box is too small to display the full name, again, you can expand the size of the dialog box by dragging one of its edges.

- Click on the Run button, which will open up a standard file save dialog box, and then you'll need to create/choose the directory where you want to store the output files for this preprocessing using the Select Folder button. Although the output will be automatically loaded into QGIS, the underlying data files will be saved to this location.

- After you've selected the directory, the preprocessing will start. The preprocessing tab will go gray; and, if you switch to the main QGIS screen, there is a progress bar for the Semi-Automatic Plug-In at the top. **Note:** Ensure that your computer does not go into hibernation, as the processing will stop.

- When the preprocessing is complete, the progress bar on the main QGIS screen will vanish, a series of beeps will be heard, and the dialog box will no longer be grayed out. The main screen will show a black-and-white image, and each of the spectral bands will be listed in the layer panel. You can now close the SCP dialog box by clicking on the X in the top right corner.

The outcome of this step is to take the downloaded data and apply two algorithms. The first converts Landsat's Digital Numbers to calibrated

BOA reflectance for the optical bands and brightness temperature for the TIR band, while the second algorithm applies an atmospheric correction. By performing an atmospheric correction, the classification results should be improved, but it will also improve other algorithms that you might choose to apply, such as NDVI, so even if you're not undertaking a classification, this plugin can be handy.

9.6.1.3 Step Three: Creating a False-Color Composite

The next step is to create a false-color composite using wavebands 5, 4, and 3, where the different land cover types of soil, urban, vegetation, and water become visible.

To build the composite, go to the menu item Raster > Miscellaneous > "Build Virtual Raster". To select the wavebands, under Input Layers, click on the ... at end of the box, which will show all the available bands, all of which start with the letters RT to indicate they're the outputs of the preprocessing. Select the TIF file for waveband 5, followed by 4 and then 3 by clicking on the checkbox. They then need to be placed in this order, which is undertaken by clicking on the waveband and holding down the left mouse button, and dragging it to the top of the list, reaper until waveband 5 is first, 4 is second, and 3 is third; the order is important, as changing the order will change the created composite image. Click on OK next to Input Layers to return to the Build Virtual Raster dialog box.

Click the "Place each input file into a separate band" checkbox so the tick appears. Below is a heading Virtual, click on ... at the end of the box underneath, and select the option Save to File, which will take you to a standard save file dialog box. Choose the directory to save the file, give it a name such as FalseColorComposite_Ulysses, and press Save. This step is required because this action creates a virtual layer in QGIS, and if you don't save it to a file, you will keep this layer, whereas saving to the file creates an image you can reload into QGIS at a later date.

Now, click Run, and you'll see a log of the processing undertaken. When the log indicates "Algorithm 'Build virtual raster' finished", you will see a new layer named whatever you called the layer when you gave it a name to save, in our case FalseColorComposite_Ulysses, at the top of your Layers box and the main image will now be in color. Click Close to shut the Build Virtual Raster dialog box. Right-click the layer name and go to Properties, select the Histogram tab and press the Compute Histogram button. You'll notice a very high peak on the far left, and so we want to remove that from the contrast stretch. So, zoom into the bottom area of the histogram and do a contrast stretch ignoring the peak. We have used the min/max values for the Red Band 0.1856–0.400, for the Green Band 0.028–0.198, and for the Blue band 0.015–0.138. The QGIS view should look like Figure 9.5a.

(a) (b)

FIGURE 9.5
Landsat-8 data acquired over the US state of Kansas, near the city of Ulysses, on April 24, 2014, displayed in QGIS as the (a) false-color composite after preprocessing and (b) false-color composite zoomed in to show individual fields. (Data courtesy of USGS/NASA.)

If you zoom into the image, you will see an image similar to Figure 9.5b, but it will be different depending on where you zoomed in, showing many red and green circles and squares. The red circles/squares indicate the presence of vegetation in the image, and the brighter the red, the more vigorous the vegetation growth. The green circles/squares indicate areas of soil, stubble, or dead vegetation. They exist due to the irrigation method, which occurs in a circular pattern from a central point, which is why the crops grow inside circles.

If your computer is struggling to display the image, it could be because, currently, QGIS will be trying to load every spectral waveband, plus your false-color composite, which takes time. Therefore, turn off all the individual wavebands under the Layers panel by clicking on the checkbox next to each waveband to turn it off. Just leave the false-color composite to be displayed, which will speed up zooming and moving around the image.

9.6.1.4 Step Four: Choosing Classification Wavebands

The next step is to choose the wavebands to be included in the classification. Go to the menu item SCP >Band set. In the dialog box, all the bands under Single band List should appear white to show they are available. Also, in the Band Set Definition panel below, the bands B1–B7 will be listed – again, dragging the edges of the dialog box to make it bigger may be helpful. These are the bands needed for the classification; bands B10 and B11 are thermals bands that we will not use but could be included in the classification process. If any of the bands B1–B7 are not in the Band Set Definition, click on the relevant band, and then click the + button to the right of the Single band list, which will add the bands to the Band Set Definition area below.

You can now close the window by clicking on the X in top corner. Your image won't change at this point, but you will have checked that you have selected the correct Band set to allow you to undertake the next step.

You've completed all the processes required to prepare your dataset for performing a land classification. At this stage, as described in Section 7.10, it's worth saving your QGIS project so you can return to it later. Before starting the second stage, you might want to save a second version of this QGIS project, with another name, as the third stage of the practical starts again from this point.

9.6.2 Second Stage: Performing a Supervised Land Classification Using Existing Training Sites

Performing a supervised land classification requires selecting a series of training sites, referred to as regions of interest (ROIs) in QGIS, which provides spectral information about each land type for the classification process. We'll show you how to do this in stage three, but for now, we're going to use a set of training sites that have already been created. Training sites within SCP have two components, the spectral signatures and their accompanying vector layer ROI shapefile; for a land classification to work, the spectral signature must be combined with their accompanying ROI shapefile.

9.6.2.1 Step Five: Importing Spectral Signatures

The next step is to download the example spectral signature set from the associated learning resource website (https://www.playingwithrsdata. com/), where you'll find a zipped file in the Chapter 9 Practical section. Download and unzip the files into a separate directory from the prepro-cessing results.

In the SCP and Dock panel on the left side, under the Layers Panel, select the Training Input tab. Click on the "Create a new training input" button, a grey square with a yellow circle in the top right corner, which will take you to a standard file save dialog box. Here you can select the folder to save the training dataset into, and give the file a name such as Ulysses Training Dataset – just make sure it is in a different folder to the spectral signatures you have just unzipped. Click on Save, and you'll see a new layer added to your layer panel.

To load the spectral signatures, on the left of the Training Input tab, click on the "Import Spectral Signatures" button; it is towards the bottom of the buttons, and is a large blue downward pointing arrow on a graph background, and this opens a dialog box. Click on the "Open a file" icon at the end of the "Select a Vector (*.shp, *.gpkg)" box, which opens a file explorer window, and navigate to your extracted spectral signature files you created at the start of this step; select the "training.shp" file, and press Open. The "Select a Vector (*.shp, *.gpkg)" box will now be populated with that file location.

Under Vector fields, the second row of boxes will now each have the word "fld" in them. Click on the small grey downward pointing triangle at the end of each box, and select the appropriate field from the drop-down list to match the title in the first row of boxes, such that

- MC ID field has MC_ID underneath,
- MC Name field has MC_name underneath,
- C ID field has C_ID underneath, and
- C Name field has C_name underneath.

Click the Import vector button on the right side. The plugin panel will go gray, and a progress bar will appear on the main QGIS window to indicate the processing is happening. This progress bar will disappear, and the Plugin dialog box will no longer be grayed out once the processing has been completed. Close the dialog box by clicking the X in the top right corner.

The "ROI & Signature list" section of the training input tab in the "SCP & Dock" panel will now be populated with the spectral signatures of the training sites. It will show the following:

- MacroClass ID (MC ID): A unique reference number for each overall class of land cover type. There are only four land cover types for the provided spectral signatures: 10 for urban, 20 for soil, 30 for vegetation, and 40 for water.
- Class ID (C ID): These are a second level of classification beneath MacroClasses, to provide further detail; for example, vegetation can be separated into crops, forests and open spaces, or crops could be subdivided into individual crop types to monitor more precisely what is being grown. Each subclass has a unique reference within the MacroClass. However, despite a number of sub-categories already being populated by the SCP plugin, we are not going to be using any of the subclasses for this simple example.
- MacroClass Name (MC Name): A description of the class, and although it's not used during the classification, it helps distinguish between different classes.
- Type: Item type that can be R = only ROI polygon, S = only spectral signature, and RS = both ROI and spectral signature.

- Color: The color of each class will be displayed on the final classification.

We are going to undertake a series of steps to simplify the visualization of this classification and make it easier to understand:

- For each MacroClasses, click on the tiny gray downward facing triangle next to the macro class number, which will collapse the sub-categories.
- The colors chosen for the MacroClasses are done automatically by the SCP, and quite often don't make the most obvious sense, so it can help to change the colors used. Double click on the color for the Urban MacroClass; this will bring up the Select Color dialog box. Choose one of the yellows in the Basic Colors on the left side for the Urban class, and click OK. The color next to Urban will now be yellow. Repeat this action with the other MacroClasses, choosing brown for soil, green for vegetation, and blue for water. Your QGIS screen should look similar to Figure 9.6

We have simplified this exercise by only having four land cover types, but of course, it is possible to have lots of land cover types. The CORINE database described above has 44 different land cover classes. You'll be able to see these in the practical in Chapter 12.

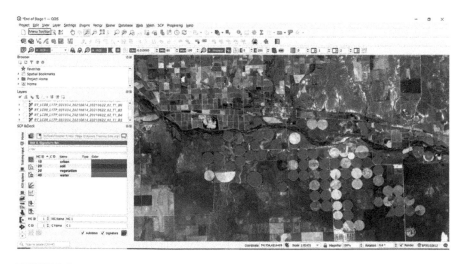

FIGURE 9.6
QGIS screen showing the imported spectral signatures after color changes. (Data courtesy of USGS/NASA.)

To classify an image effectively, you'll need to identify the range of spectral signatures within it, and this requires identifying multiple training sites for each land cover type. The more land cover types you have, the more training data will be required. Having a few training sites will still classify the image, but there may be confusion between them; for example, the spectral signature of urban sites and soil can be quite close, and without a sufficient range of training sites, urban areas may be incorrectly classified as soil. Of course, this is always partially true as urban environments contain soil, but this is about classifying the predominant feature for each area. This also explains why you may find that sometimes you see some MacroClasses/Subclasses having the same descriptions and color. Each different land type has various spectral signatures as; for example, vegetation includes forests, crops, and open spaces such as grassed areas. Also, forests include different kinds of trees with different spectral signatures, crops would include wheat and rice, and even open spaces may be grassland, moorland, or scrubland. This is also the same for urban, soil, the water land types.

9.6.2.2 *Step Six: Classification Algorithm and Preview*

Several algorithms are available to undertake the classification; for this practical, we will be using the "Minimum Distance" classification.

Go to the menu item SCP > Band processing > Classification and will open up the classification dialog box. Under the Algorithm section, click the downward-facing gray triangle to show the drop-down classification algorithm options. The three choices are "Minimum Distance", "Maximum Likelihood", and "Spectral Angle Mapping" with each method using a different approach for separating the surface types with the following:

- Minimum Distance: Calculates the shortest distance between each pixel and the mean vector for each class, with the closest distance deciding which class is chosen
- Maximum Likelihood: Assumes that the statistics for each class are normally distributed and then calculates the probability that a given pixel belongs to a specific class
- Spectral Angle Mapping: Determines the spectral similarity by treating the spectra as vectors in space, and then calculating the angles between them

Select "Minimum Distance" from the drop-down menu. Also, as we're using the MacroClass ID (MC ID), ensure that "Use MC ID" checkbox has a tick in it, and if it doesn't, select this option by clicking on it. Having selected these options, close this dialog box by clicking on the X in the top right corner.

A classification of a whole Landsat image does take some time, particularly if you have a lot of classes, so the SCP can undertake a classification over a small subset area to see how it works; this is particularly important when you're setting up your training sites as we'll explain in stage three. This is known as a Classification Preview. To undertake the preview, firstly, zoom into your image until you can see some of the individual circles and squares. Next, on the SCP toolbar, there is a "Activate Classification Preview pointer" icon that is a quartered square with red, blue, yellow and green sections overlaid with a white plus sign, click this button and as you move your cursor back over the view you will see that it has changed to a +. Choose somewhere on your view, and then left-click with your mouse, which will activate the classifier, processing a small area of the image with your chosen point at the center. This processing will overlay the classification onto the view, similar to Figure 9.7a; showing areas of the different classes, and you can zoom in to see more details. By displaying the classification and false-color composite layers, you can switch between the two easily by turning one layer on and off, and you can see how the two layers match each other. In our case, Figure 9.7b is the false-color composite, and you can see fields with vegetation, bare soil, small areas of water, and some urban buildings; it also shows some areas as urban that look more like soil as we described above.

You can repeat the classification preview as often as you like by clicking the "Activate Classification Preview pointer" icon and choosing a point on your view. Although, as it creates a new layer each time under the Class_temp_group, you may want to delete any unwanted layers to avoid confusion later.

You'll have noticed by now that QGIS needs to save data files every time you create new layers because the QGIS project itself does not hold the original data, just the metadata to support the visualization. When you create a layer, such as the classification preview layer, where QGIS does not request a location for the data file, it won't save the data even if you save the project. Therefore, if you want to save any temporary layer, right-click on that layer and then select "Save As...."

9.6.2.3 *Step Seven: Whole Scene Classification*

When you're satisfied with the classification preview testing, you can perform the full scene classification by going to the menu item SCP > Band processing > Classification and clicking Run in the dialog box. It will first take you to a file save dialog box to select the directory to save the GeoTIFF classification image into, and give it a name, such as Ulysses Full Classification. Clicking Save starts the full classification.

It takes a short period to do the processing, and as usual, the dialog box will be grayed out during the process, and a bar at the top of the main

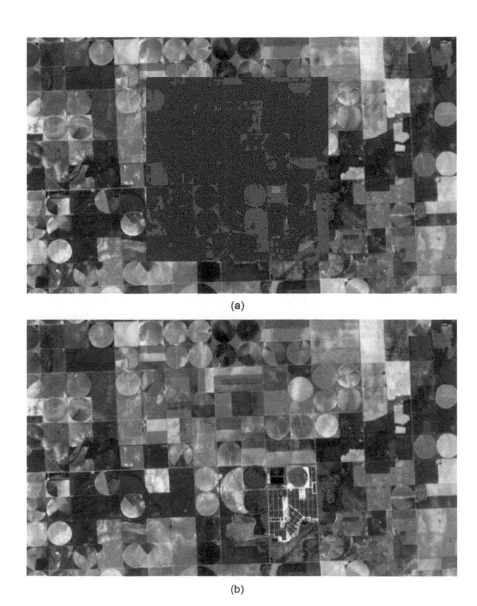

(a)

(b)

FIGURE 9.7
Landsat-8 scene as the (a) classification applied using the imported spectral signatures and (b) associated false-color composite. (Data courtesy of USGS/NASA.)

window will show progress. Once the processing has been completed, you can then close the dialog box. The top layer will now be the full classification; a classified layer should be classified using the training data loaded in.

This stage should have demonstrated how the land classification process works, using the provided spectral signature file for this part of the US state of Kansas. However, suppose you want to classify a different area of the world, or even different parts of Kansas. In that case, you'll need to create your own training sites relevant to that image/geographical location, which is covered in the next stage of the practical.

9.6.3 Third Stage: Performing a Supervised Land Classification with Your Own Training Sites

This stage will create a land classification for the same area of Kansas, but this time, you will create your own training sites. We start this part of the practical with the processed data, but without any land classifications. Thus, go back to the copy of the project you saved at the end of the first stage. Use the menu item Project > Open; navigate to the saved project, select it and press Open.

You should have your false-color composite as the top layer and the classification wavebands from the first stage in your layers – ideally with the layers turned off.

9.6.3.1 Step Eight: Creating a Pseudo-True-Color Composite

The first part of this stage is to create a pseudo-true-color composite, in the same way as for the false-color composite in step three. Go to the menu item Raster > Miscellaneous > Build Virtual Raster.

Under Input Layers, click on the ... at end of the box to show the available bands. Select the files for waveband 4, followed by 3 and then 2, and ensure that they are placed in this order at the top. Click on the OK next to Input Layers to return to the Build Virtual Raster dialog box.

Click on the "Place each input file into a separate band" checkbox to show the tick inside the checkbox. Below there is a heading Virtual, click on ... at the end of the box underneath, and select the option to Save to File, which will take you to a standard save file dialog box. Choose the directory to save the file, give it a name such as TrueColorComposite_Ulysses, and press Save.

Now, click Run and the log of actions until "Algorithm 'Build virtual raster' finished". Click Close to shut the dialog box. A new layer, TrueColorComposite_Ulysses, will be on top, and you should have an image that looks like Figure 9.8a.

You've now created both the false color and pseudo-true-color composites within QGIS. If you use the checkboxes to turn one of them off and on, you'll be able to see the different landscape features in both true and false color; this works particularly well when the view is zoomed in.

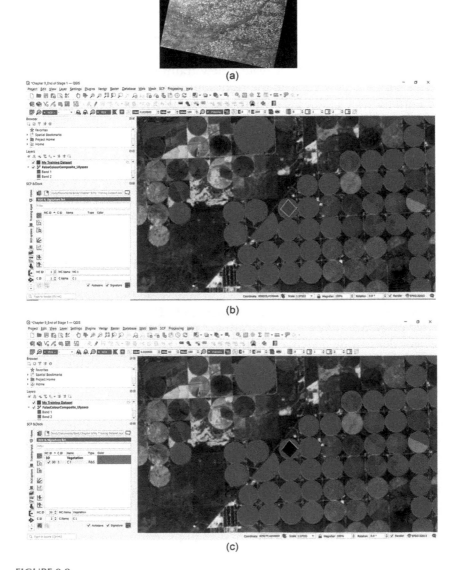

FIGURE 9.8
Landsat-8 scene as the (a) pseudo-true-color composite, (b) QGIS screen zoomed in to show false-color composite with ROI polygon selected and (c) QGIS screen zoomed in false-color composite with training data polygon classified. (Data courtesy of USGS/NASA.)

The pseudo-true-color composite should help identify features to classify them, potentially alongside other tools such as Google Earth and Google Maps.

9.6.3.2 Step Nine: Identifying and Selecting Your Own Training Sites

First, you'll need to create a layer to save the ROIs within. Select the Training Input tab in the SCP and Dock panel on the left side under the Layers Panel. Click on the "Create a new training input" button, a gray square with a yellow circle in the right top corner; this will take you to a standard file save dialog box, where you select the folder to save the training dataset into, and give the file a name such as My Training Dataset.

The next stage is to create the training sites, and for this initial step, we will identify the same four land classifications we used in stage 2: vegetation, soil, urban, and water. As demonstrated in the ESA video, vegetation appears red in this false-color composite. Turn off the true-color composite layer created in the previous step, and zoom into a cluster of red circles in the top right corner of the view, which are dominated by green vegetation. Although, there will still be variability within these areas. You'll find it easier to be zoomed in for this part of the practical.

To create a training site, click on the "Create an ROI polygon" button on the toolbar, it's an orange square with part of the right side missing and some red dots and x's on it, and when you move onto the image the cursor will be a + and it will have a number next to it. This is the NDVI number of this pixel and if you move the cursor around the image, the number changes depending on what is in the pixel.

Move the cursor to one of the bright red circles and left click towards the top of the circle but inside the circumference, then move to the right side of the circle inside the circumference and left click again. You will see a white line joining the two points you clicked – there are also red crosses at the end of each line, but you can't see them because of the red circle being used as the ROI. Repeat this process for a point at the bottom of the circle, and then on the left side, always inside the circumference, and finally go back to your starting point at the top of the circle, but this time right-click to close the polygon. You should have a semi-transparent orange polygon inside the red circle, similar to Figure 9.8b. Although we created four points, you can have as many sides to your polygon as you prefer before you close it. Still, you need to ensure that all the points are over the same type of land, or in this case, within the circle's circumference, to ensure the correct average spectral signature is calculated.

Having identified a training ROI polygon, it needs to be classified, and this is performed under the "ROI & Signature list" section on the Training Input tab. Go to the bottom of the tab, and give it a unique number in the

MC ID field - to keep the same classification numbering/grouping system that was used in the second stage, type "30" for the ID, and then type Vegetation in the MC Name. Finally, click on the "Save temporary ROI to training input" button in the bottom right corner of the tab; it is a gray rectangle with yellow and orange squares underneath a black background.

Once SCP has extracted the average spectral signature for your selected polygon, you'll see vegetation in the ROI list with the MC ID of 30. The image will now show a black area where your semi-transparent orange polygon was, as shown in Figure 9.8c. QGIS will automatically assign the spectral signature a color for how the feature will appear on the image; however, you can change it as described in step five.

Now repeat this to identify training sites for the soil, urban, and water classes using the false-color composite:

- Soil: This predominately appears as shades of gray, with unplanted circles or rectangles, alongside the regions that aren't cultivated. If you want to select precisely the same site we've used, you can use the coordinate reference "809142,4085966" shown at the bottom of the figure by directly entering the coordinates given here into that box. Select your soil training site manually by drawing a polygon as above, although this time you should see the red x's as you select the sides of the polygon. Once you have closed the polygon with the right click, call it "Soil" in the MC Name, give it the MC ID of "20", and then click on the "Save temporary ROI to training input" button. Change the color of the classes if you want, and you should have something that looks similar to Figure 9.9a.
- Urban: We've selected the city of Hugoton with the coordinate reference "824652,4120472", which shows up as pale blue areas. Select your urban training site as a polygon encompassing a pale blue area in the city's center as in Figure 9.9b. Call it "Urban" in the MC Name, give it the MC ID of "10", and change the color if you want.
- Water: The Arkansas River flows across the middle of the top third of the image, and we have chosen a body of water just off the river. This is a fairly small area, so you must select your polygon carefully to ensure it remains within the black pixels. In Figure 9.9c, our chosen site is at coordinate reference "747271,4215089". Select your water training site, again using a polygon inside the water body. Call it "Water" in the MC Name, give it the MC ID of "40", and change color if you want.

Having completed these four training sites, you'll now have a basic set of land classifications.

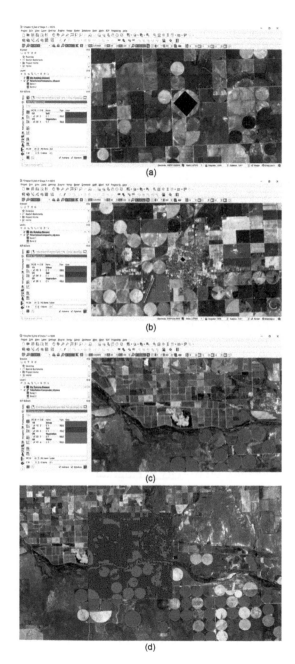

FIGURE 9.9
Landsat-8 scene zoomed in to show the training region-of-interest sites picked for (a) soil, (b) urban, and (c) water, plus the (d) classification results for a subarea. (Data courtesy of USGS/NASA.)

9.6.3.3 *Step Eleven: Classification Algorithm and Preview*

To set up the classification for the preview, firstly, repeat step six to determine the classification algorithm; by going to menu item SCP>Band Processing>Classification from the top menu, select the "Minimum Distance" option from the Algorithm drop-down menu, and ensure that the MC ID checkbox has the tick in it. Close the dialog box, by clicking on the X in the top right corner.

You can now perform a classification preview on the part of the view by clicking on the "Activate Classification Preview pointer" button on the SCP toolbar; it is a quartered square with red, blue, yellow, and green sections overlaid with a white plus sign. As you move your cursor back over the image view, you will see that it has changed to a +, and clicking somewhere on the main view will give you a preview of your land classification; see Figure 9.9d.

While you'll have a land classification, it's almost certain that it won't be a perfect match as we've chosen only a single training site for each land type. Therefore, rivers may be classified as urban areas, vegetation classified as water, and so on. This result is primarily because the classification doesn't have enough spectral signatures; as previously discussed, the signatures can vary significantly for a single classification type. Therefore, the accuracy of the classification can be improved/adjusted, by creating several training sites for each classification type so that the spectral signature variations are captured.

At this stage, you might find the pseudo-true-color composite, and other reference sources such as Google Maps and Google Earth, useful to help identify different features. As in the second stage, it might be easier to have all the groupings of the same land cover type with the same or similar colors so that it's easy to see the different land classifications.

You would need to have between 10 and 15 training sites for each of your classes to get a reasonably accurate classification, although even this is a minimal number. For classifiers we develop, it is not usual to have 10,000 training sites or more.

9.6.3.4 *Step Ten: Whole Scene Classification*

Once you've got a classification with sufficient training sites to produce a good level of accuracy, you can perform the full scene classification by going to the menu SCP > Band processing > Classification, and in the dialog box click Run. This will first take you to a file save dialog box to select the directory to save the tif classification image into, and give it a name, such as Ulysses Full Classification. Clicking Save starts the full classification, and when the processing has completed, you will have undertaken a land classification using training data you generated yourself.

NOTE: the classification will not be hugely accurate with just four training sites, but you can refine this by adding as many as you like. While this is just over Ulysses, you can now follow this process to undertake land classifications anywhere in the world.

9.7 Summary

This chapter has focused on using time-series and land classification techniques to monitor large geographical landscape evolution; we've particularly highlighted the optical indices used in this type of remote sensing, with the theory focusing on how these are calculated.

The real-world applications have used changes in land classification over short- and long-term time frames to monitor landscape, agriculture, forests, and fires. While the practical exercise is a bit complicated and challenging, it gives you an impressive skill set to assess land type and changes over any area of the world. Therefore, this approach enables you to investigate the changes in your own local area, looking at how much green space exists, where landscape usage has altered, or whether forests have expanded or contracted. You've got lots of exciting possibilities to research.

9.8 Online Resources

- ALOS-2 PALSAR-2 sample data, used for Figure 9.3: https://www.eorc.jaxa.jp/ALOS-2/en/doc/sam_index.htm
- Associated learning resource website: https://playingwithrsdata.com/
- Centre for Environmental Data Archival (CEDA), MTCI data: https://catalogue.ceda.ac.uk/uuid/9ed8d70ffde5b757691f3d8124f13148
- Copernicus Atmosphere Monitoring Service: https://www.gmes-atmosphere.eu/
- Copernicus Land Monitoring Service: https://land.copernicus.eu/
- CORINE Land Cover data sets: https://land.copernicus.eu/pan-european/corine-land-cover

- Digital Earth Africa https://www.digitalearthafrica.org/
- Harmonized Landsat Sentinel-2 https://www.earthdata.nasa. gov/esds/harmonized-landsat-sentinel-2
- ICESAT-GLAS: https://nsidc.org/data/icesat/
- Landsat news page: https://www.usgs.gov/landsat-missions/news
- MYD14 Thermal Anomalies and Fire products are detailed at: https://lpdaac.usgs.gov/products/modis_products_table/myd14
- NASA Open Data Portal: https://data.nasa.gov/
- Semi-Automatic Plugin (SCP) support page: https://fromgistors. blogspot.com/.
- Sen2Like: https://github.com/senbox-org/sen2like

9.9 Key Terms

- Land use and land cover: Two ways to describe the land surface.
- PCA: An example of an unsupervised classification where the data are transformed into a form that defines the linkages in terms of variance rather than distance, before the classification is applied.
- Root-mean-squared error: Statistical value often used to represent an overall positional error for an image.
- Supervised classification: Where a computer algorithm performs the classification based, and then the user interprets the results.
- Unsupervised classification: Where the user defines a library of training sites, which a computer uses to then assign a classification to any pixel grouping with the same characteristics.
- Vegetation indices: Use waveband ratios to emphasize the green photosynthetically active vegetation.

References

Buttner, L. 2014. CORINE land cover and land cover change products. In *Land Use and Land Cover Mapping in Europe: Practices and Trends*, eds. I. Manakos and M. Braun, 57–74. Berlin, Heidelberg: Springer-Verlag.

Byrne, G. F., P. F. Crapper and K. K. Mayo. 1980. Monitoring land cover change by principal component analysis of multitemporal Landsat data. *Remote Sens Environ* 10:175–184.

Chen, C. and H. Mcnairn. 2006. A neural network integrated approach for rice crop monitoring. *Int J Remote Sens* 27(7):1367–1393.

Clevers, J. G. P. W. 2014. Beyond NDVI: Extraction of biophysical variables from remote sensing imagery. In *Land Use and Land Cover Mapping in Europe: Practices and Trends*, eds. I. Manakos and M. Braun, 363–381. Berlin, Heidelberg: Springer-Verlag.

Curran, P. J. 1989. Remote sensing of foliar chemistry. *Remote Sens Environ* 29:271–278.

Deering, D. W. and R. H. Haas. 1980. Using Landsat digital data for estimating-green biomass. NASA Technical Memorandum. Available at http://ntrs.nasa.gov/archive/nasa/casi.ntrs.nasa.gov/19800024311.pdf (accessed April 17, 2015).

Duane Nellis, M., K. P. Price and D. Lundquist. 2009. Remote sensing of cropland agriculture. In *The SAGE Handbook of Remote Sensing*, eds. T. A. Warner, M. Duane Nellis and G. M. Foody, 368–380, London: SAGE Publications.

Gao, B.-C. 1996. NDWI—A normalized difference water index for remote sensing of vegetation liquid water from space. *Remote Sens Environ* 58:257–266.

Henrich, V., G. Krauss, C. Götze and C. Sandow. 2012. IDB—www.indexdata base.de, Entwicklung einer Datenbank für Fernerkundungsindizes. AK Fernerkundung, Bochum. Available at http://www.lap.uni-bonn.de/pub likationen/posterordner/henrich_et_al_2012 (accessed April 17, 2015).

Lucas, R. M., X. Xiao, S. Hagen and S. Frolking. 2002. Evaluating TERRA-1 MODIS data for discrimination of tropical secondary forest regeneration stages in the Brazilian Legal Amazon. *Geophys Res Lett* 29(8):1200. Available at http://doi.org/10.1029/2001GL013375.

Richards, J. A. 1984. Thematic mapping from multitemporal image data using the principal components transformation. *Remote Sens Environ* 16(1):35–46.

Richardson, P., P. J. Astrand and P. Loudjani. 2014. The CAP fits. *GeoConnexion International Magazine*, November/December: 22–24.

Thomson, A. G. 1992. A multi-temporal comparison of two similar Landsat Thematic Mapper images of upland North Wales, UK. *Int J Remote Sens* 13(5):947–955.

Wang, Y., L. L. Hess, S. Filoso and J. M. Melack. 1995. Understanding the radar backscattering from flooded and non-flooded Amazonian forests: Results from canopy backscatter modeling. *Remote Sens Environ* 54(3):324–332.

10

Inland Waters and the Water Cycle

The terrestrial water cycle describes how rivers, lakes, and wetlands gain and lose water. Water comes down from the atmosphere in the form of precipitation; globally, around two-thirds of this goes back into the atmosphere through evaporation. The rest restocks groundwater and provides surface and subsurface runoff, which flows down rivers into lakes and reservoirs to provide freshwater. This freshwater is critical for human life and nature but only accounts for around 2.5% of the total water on Earth (Vörösmarty, 2009).

As the human population increases, the pressures on the terrestrial water system increase; for example, the threefold increase of the global population during the 20th century has triggered a simultaneous sixfold increase in water use (FAO, 2015). In addition, water bodies are affected by longer-term drivers such as climate and land use change.

Understanding the dynamics of the terrestrial water cycle and its long-term drivers is vital to understanding freshwater behavior and sensitivity in rivers, lakes, reservoirs, and groundwater. Traditional approaches to monitoring water are primarily based on in situ sensors, such as water gauges, but the costs have resulted in geographically and temporally sparse data. Satellite remote sensing can provide a global approach with multiple parameters to support water cycle monitoring. This approach is of particular benefit in remote areas, or areas with poor or nonexistent in situ monitoring networks, as the changes can be determined without the need to visit the physical locations regularly. Where in situ gauges and other data sets are available, remote sensing can complement and extend these data sets by providing more significant insights. The nature of Earth Observation (EO) archives and data sets also offers possibilities to study both shorter- and longer-term drivers over time.

10.1 Optical and Thermal Data for Inland Waters

Water tends to absorb most of the incident electromagnetic (EM) radiation, so it has a lower reflectance in the visible and infrared (IR) wavelengths than the surrounding features. Therefore, a key use for optical data is determining water extent. McFeeters (1996) developed a Normalized

DOI: 10.1201/9781003272274-10

Difference Water Index (NDWI) using the green to near-IR (NIR) wave-band ratio:

$$NDWI = \frac{(Green - NIR)}{(Green + NIR)} \qquad (10.1)$$

The input to this waveband ratio should be the surface reflectance, rather than raw uncalibrated data, and water features should have positive values while vegetation and soil usually have zero or negative values (McFeeters, 1996). However, artificial features can appear similar to water bodies. In addition, the contrast with the surrounding soil and vegetation is reduced when the water is turbid because of the material suspended in it. This approach was expanded by Xu (2006) who separated land from water using a green-to-MIR ratio, which has the advantage that water will remain dark in the MIR, even with high concentrations of material suspended in it; man-made surfaces have a negative value for this ratio:

Modified Normalized Difference Water Index

$$(MNDWI) = \frac{(Green - MIR)}{(Green + MIR)}. \qquad (10.2)$$

The second key element of optical inland water monitoring is the water quality, that is, the amount of organic and inorganic material suspended in the water and water temperature. The suspended material includes phytoplankton that forms the basis of the aquatic food chain and strongly influences the water quality in both the oceans and freshwater bodies. Monitoring suspended material in water is undertaken through the optical signature and is known as ocean color as it started in the open ocean; the theory will be discussed in depth in Chapter 11. Lake surface water temperature (LSWT, often written as LST, but we've used that abbreviation for land surface temperature) is derived from thermal IR (TIR) wavebands because the passive microwave data have too coarse spatial resolution. Figure 10.1a shows the Landsat scenes that cover Lake Victoria in Africa that straddles the borders of Kenya, Tanzania, and Uganda. The LWST for the lake is shown in Figure 10.1b using 1-km spatial resolution data from the European Space Agency (ESA) ARC Lake project based on the Along Track Scanning Radiometer (ATSR) missions, while Figure 10.1c has a better lake shape showing the ocean color–derived water quality using the Medium Resolution Imaging Spectrometer (MERIS) Full-Resolution Chlor-a product with 300-m spatial resolution. The limited spatial resolution of optical sensors and the difficulties caused by the complexity of the top-of-atmosphere reflectance signal has been improved with the launch of hyperspectral missions and the long-term data archive from

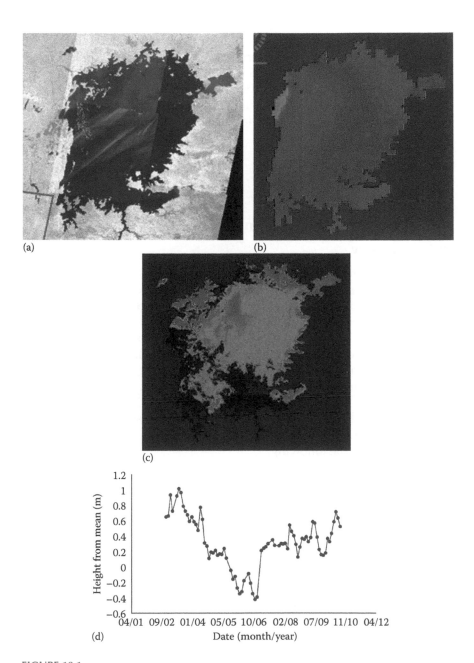

(a)   (b)

(c)

(d)

FIGURE 10.1
Lake Victoria examples including (a) Landsat Thematic Mapper mosaic for June 2009, (b) lake surface water temperature from ARC-Lake, (c) March 2012 chlorophyll-a product created using the eutrophic lakes processor applied to a MERIS FR image, and (d) lake-level data from the European Space Agency River and Lake project. (Data courtesy of the named projects alongside ESA/NASA/USGS.)

the Landsat satellite where the upgraded sensors (Operational Land Imager [OLI] and OLI-2) have improved radiometric spectral capabilities (e.g., Palmer et al., 2015). The launch of the Sentinel-2 missions carrying the MultiSpectral Instrument (MSI) and Sentinel-3 satellites carrying the Ocean and Land Color Instrument (OLCI) and Sea and Land Surface Temperature Radiometer (SLSTR) have significantly increased capabilities and data availability further.

10.2 Microwave Data for Monitoring the Water Cycle

10.2.1 Altimetry

Altimetry uses microwave energy in Ku, C, S, and Ka frequency bands. However, there are challenges in using microwave energy as the available bandwidth is determined by international regulations because of potential conflict with other activities, such as Wi-Fi which limits the emitted power. The C- and S-bands are often used in combination with the Ku-band owing to atmospheric sensitivity, as they help estimate the ionospheric delay caused by the charged particles in the upper atmosphere. The Ka-band offers better observations of ice, inland waters, and coastal zones because a larger bandwidth is possible but is sensitive to rain. The operation of traditional altimeters, which are pulse limited, is termed low-resolution mode (LRM), and examples of satellites using this type of operation include the following:

- Jason-1, Jason-2, and Jason-3 operating at the Ku frequency were followed by the Sentinel-6 Michael Freilich Mission also called the Jason-CS (Jason Continuity of Service) mission.
- ERS-1, ERS-2, and Envisat Radar Altimeter-2 (RA-2) having operated in the C- and S-bands
- Saral/AltiKa operating in the Ka-band

Altimetry is useful for inland waters because it's possible to measure the height of water levels as the land is a relatively poor reflector of Ku/Ka-band energy compared with inland water (Berry et al., 2005). However, there are difficulties when using altimetry over small water bodies as the data needs to be filtered to ensure that only reflections from the water surface are included (Crétaux et al., 2011). This filtering is because the footprint of the energy from each pulse covers an area of approximately 2–5 km in diameter, which depends on the roughness of the surface; with rougher areas, more reflected energy can find its way back to the satellite sensor receiver as shown in Figure 10.2.

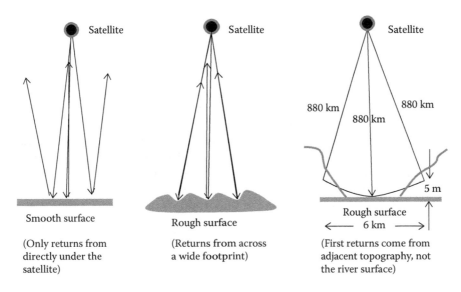

Satellite   Satellite   Satellite

880 km   880 km

880 km

5 m

Smooth surface   Rough surface   Rough surface

←—— 6 km ——→

(Only returns from directly under the satellite)

(Returns from across a wide footprint)

(First returns come from adjacent topography, not the river surface)

FIGURE 10.2
Altimetry.

CryoSat-2 carries the Synthetic Aperture Radar (SAR) Interferometric Radar Altimeter (SIRAL), which is a single-frequency Ku-band instrument. It includes the traditional LRM, alongside an altimetric SAR mode offering higher-resolution measurements and an altimetric SAR Interferometric (SARIn) Mode, with the two receiving antennas forming an interferometer in the cross-track direction, able to provide improved measurement locations over sloped surfaces; both of these latter modes have an improved spatial resolution compared with LRM. When launched in 2010, its original aim was to measure the thickness and circumference of the polar ice sheets and sea-ice cover; after adjustments to its data collection method, it can now be used for new applications such as mapping terrestrial water bodies.

10.2.2 Passive Radiometry

Microwave radiometry can provide precipitation estimates, such as rainfall, with microwave measurements providing information about the insides of clouds, while IR radiation complements this with information about the top of clouds. Microwave radiometers separate the radiation originating from the Earth's surface from that arising from precipitation, with estimates being more straightforward over the ocean because emissions from this surface are strongly polarized, while emissions from raindrops are depolarized. Thus, precipitation is distinguished using vertically and horizontally polarized radiation measurements.

10.3 Inland Water Applications

10.3.1 Water Cycle and Wetlands

Monitoring the overall terrestrial water cycle looks at the big picture, like the Land Use and Land Cover mapping from the previous chapter. As such, understanding the full water cycle requires combining EO data with other data sets and models because it includes land surface interactions, heat fluxes, and soil surface moisture. The overall water cycle includes snow and ice, soil moisture, groundwater, and within water bodies, which are covered in their relevant sections below.

Rainfall is a critical component of the overall water cycle, and there have been several precipitation-focused missions. An example is the Tropical Rainfall Measurement Mission (TRMM), which operated between 1997 and 2015 and carried both passive and active sensors, including a multichannel dual-polarized passive microwave radiometer with a 14-GHz rain radar and 6-waveband visible/NIR instrument. To capture the rain variability during the day, the orbit was inclined at 35°; hence, the overpasses occur at different local times on successive days. The follow-on is the Global Precipitation Measurement (GPM) mission, an international network of satellites to provide the next-generation global observations of rain and snow. The GPM Core Observatory was launched in February 2014 with a dual-frequency Ku/Ka-band precipitation radar and a multichannel GPM microwave imager. Further information is available on the National Aeronautics and Space Administration (NASA) GPM missions' website (https://gpm.nasa.gov/).

Wetlands combine vegetation and open water, where water covers the soil. They provide a disproportionately high amount of ecosystem services for their area compared with other ecosystems, but growing pressures such as population growth and climate change are causing strong impacts. Inland wetlands occur on floodplains alongside rivers, on the margins of lakes, and in other low-lying areas where the groundwater intercepts the soil surface or where precipitation sufficiently saturates the soil. Coastal (or tidal) wetlands occur where freshwater and seawater mix and act as coastal erosion and inundation barriers. These areas include tidal salt marshes, swamps, and mangroves; mangroves are discussed further in this chapter.

Small changes in wetland hydrology can potentially result in significant changes in wetland function and extent. Wetland mapping can involve using Digital Elevation Model (DEM) data to create data layers related to hydrologic conditions and approaches such as NDWI and SAR imagery to map open water; see an example in Figure 10.3 that indicates the loss of wetlands in Uganda from a water extent map

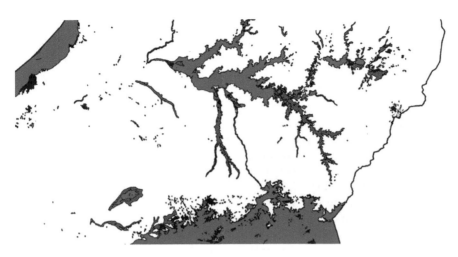

FIGURE 10.3
Loss of wetlands in Uganda. Blue – September 2018 water extent map automatically generated using Copernicus Sentinel-1. Red, comparison with the 2005 map from the Uganda
Bureau of Statistics. (Data courtesy of Copernicus/European Space Agency with the processing undertaken within the DFMS project.)

generated using Copernicus Sentinel-1 data. The Drought and Flood
Mitigation Service project was funded by the UK Space Agency and led
by the RHEA Group, in cooperation with the Government of Uganda
(Lavender et al., 2017).

Techniques for mapping vegetation types, including the change over
time, with both supervised and unsupervised classifications as described
in Chapter 9.

10.3.2 Soil Moisture Monitoring

As well as contributing to the overall water cycle, soil moisture plays a
critical role in seed germination and plant growth, key parameters for
irrigation and crop yield prediction. In addition, soil moisture products
have the potential to provide early flood detection for large-scale events,
in particular, where in situ gauged networks are unavailable.

There have been a series of missions focused on detecting soil moisture
using passive microwave signals. The results are most accurate when there
is low vegetation cover, as vegetation will attenuate the signal. The first was
the ESA Soil Moisture and Ocean Salinity (SMOS) mission, launched in
November 2009, with the dual aim of understanding soil moisture and ocean
salinity. It's a passive microwave satellite operating in the L band and has
the dual aim of understanding the water cycle processing on land and in the
oceans. As described in Section 3.2, it exploits the interferometry principle

with 69 receivers that have their signals combined to a hexagon-like shape approximately 1,000 km across with a spatial resolution of around 35 km at the center of the field of view.

The second soil moisture instrument was NASA's Aquarius, launched in June 2011 onboard the Argentinean spacecraft Aquarius/Satellite de Aplicaciones Científicas. It was originally designed to measure ocean salinity but also produces global soil moisture maps. SMOS and Aquarius have a coarser resolution than the third mission, which is NASA's Soil Moisture Active Passive (SMAP) satellite, designed explicitly for soil moisture. Aquarius failed in November 2014 when it suffered a hardware failure.

Launched on January 31, 2015, SMAP measures the amount of water in the top 0.05 m of the soil. It carries an L-band radiometer, combined with an L-band SAR instrument, providing information sensitive to both vegetation and soil moisture. The radiometer has a high soil moisture measurement accuracy but has a spatial resolution of only 40 km, whereas the SAR instrument has a much higher spatial resolution of 10 km, but with lower soil moisture measurement sensitivity. Combining the observations provides measurements of soil moisture at 10 km spatial resolution and a freeze/thaw ground state at 3 km resolution. In July 2015, 208 days after launch, the radar portion of SMAP failed when its orbit overlapped the South Atlantic Anomaly – an area of great significance to satellites as they are exposed to strong radiation, caused by trapped protons in the inner Van Allen belt. Therefore, SMAP has been combined with Sentinel-1, a replacement active microwave dataset, to produce soil moisture estimates at 3 km resolution. In addition, adding GPM data to the SMAP soil moisture algorithm provides a more robust approach for assessing the times and locations of rainfall events so that the SMAP mission can correctly interpret the soil moisture data.

The ESA-funded Soil Moisture Climate Change Initiative (CCI) project (https://climate.esa.int/en/projects/soil-moisture/) brings together soil moisture-focused missions alongside additional passive and active microwave missions to provide users with a continuous dataset, from 1979 onward, in support of scientific studies and applications. The full range of CCI datasets can be accessed from ESA's climate website (https://climate.esa.int/en/). Future soil moisture satellites include the dedicated US and Indian NASA–Indian Space Research Organization Synthetic Aperture Radar (NISAR) and the German Tandem-L (Moreira et al., 2015) missions that will be able to provide soil moisture retrievals at a higher spatial resolution and will open a market for new applications. The NISAR mission will be a dual-frequency (L- and S-band) SAR to understand natural processes such as ecosystem disturbances, ice-sheet collapse, and natural hazards such as earthquakes, tsunamis, volcanoes, and landslides. Tandem-L is based on using two radar satellites operating in L-band, which will help it resolve the vertical structure of vegetation, ice, and surface deformation.

10.3.3 Lakes, Rivers, and Reservoirs

Rivers, lakes, and reservoirs are the holders of the majority of freshwater in the world and are key conduits for the terrestrial water cycle. Therefore, understanding the quality and volume of water in these elements is critical. Remote sensing uses boundaries and water heights of lakes, rivers, and reservoirs to help determine these parameters.

Satellite altimetry uses several different retracking methods to determine the leading edge of the radar return signal, which is used to calculate the water height of the water body. The decision on which method to use depends on the surface of the ground being measured, and the choice can result in errors from the order of several tens of centimeters (Birkett and Beckley, 2010) up to several meters (Frappart et al., 2006) in the final estimated range and hence water level.

Constructing a lake or river water-level time series requires careful data editing and filtering of the altimetry data to ensure that only waveforms where the leading edge can be determined accurately are selected (Crétaux et al., 2011). Therefore, a vital ingredient for smaller lakes includes an accurate inland water location mask that allows short segments of data containing the water surface to be selected from the land.

River- and lake-level data are available from several sources, including the following:

- The AVISO website (https://www.aviso.altimetry.fr/en/home.html), which contains altimetry data for marine and hydrological applications
- Copernicus Global Land Service water-level products (https://land.copernicus.eu/global/products/wl)
- HYDROLARE project (http://hydrolare.net/) that provides water-level variations for approximately 150 global lakes and reservoirs
- US Department of Agriculture's Foreign Agricultural Service Crop Explorer portal (https://ipad.fas.usda.gov/cropexplorer/Default.aspx)

Changes in lake areas are clearly visible through time-series images; for example, Figure 10.4 shows the Dead Sea that straddles the borders of Israel, the West Bank, and Jordan that has dropped by more than 40 m in water-level height since the 1950s. The primary cause is the diversion of river water, which used to flow into the sea is used for a growing population in the area, industry, and irrigation. Figure 10.4a and b shows that the larger Northern area is shrinking, while the smaller Southern area has been subdivided to support salt production.

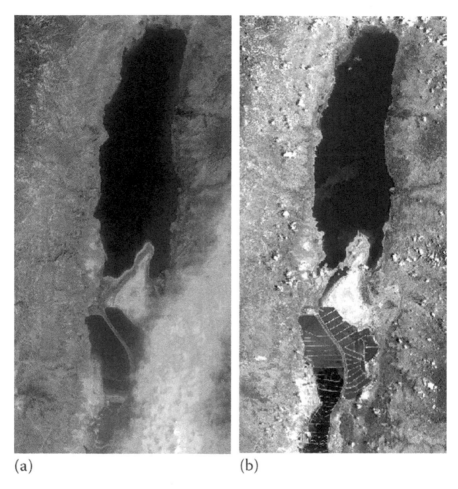

(a)   (b)

FIGURE 10.4
Landsat imagery showing the change in the Dead Sea, through pseudo-true-color composites of (a) Landsat-5 MSS data acquired in 1984 and (b) Landsat-8 OLI in 2014. (Data courtesy of NASA/USGS.)

Eutrophication is when lakes and rivers become enriched with nutrients, often through agricultural land surface runoff, causing harmful algal blooms to grow. Blooms may use up the oxygen in the river/lake, causing the death of marine inhabitants, blocking sunlight from underwater photosynthetic plants, and causing toxic effects to enter the marine food chain. The techniques for monitoring this issue by remote sensing are described in more detail in Chapter 11. However, there are programs such as the CoastWatch Great Lakes program (https://coastwatch.noaa. gov/cw/index.html), which is part of a National Oceanic and Atmospheric

Administration (NOAA) program, to produce and deliver environmental data in near-real time (NRT) and as retrospective monitoring.

10.3.4 Flood Mapping

Floods are the costliest type of natural disaster, globally accounting for approximately one-third of all reported events and associated economic losses (Wolfgang Kron, 2005), and with climate change leading to increased storminess and sea-level rise (Folland et al., 2001), the impacts will increase.

During an episode of flooding, there's an urgent need for both local and regional mapping products to help inform management decisions. These products need to be delivered in an NRT timescale, with both the extent and depth of floodwaters important for determining accessibility and flood volumes. As time-critical reliable information is needed under rainy or at least cloudy conditions, microwave data are the best solution for the initial mapping activity, with altimetry and SAR offering the potential. SAR high spatial resolution data has predominately only been offered commercially, so coverage was an issue unless the satellite was tasked. However, with newer commercial operators, constellations are becoming the focus, reducing previously relatively long revisit times. Satellite missions such as Capella Space's and ICEYE's X-band SAR satellites, COSMO-SkyMed, Sentinel-1, and TerraSAR-X/Tandem-X have more than one satellite in orbit.

Space agencies and commercial operators work together under the auspices of the International Charter Space and Major Disasters (https://www.disasterscharter.org/). When a Charter member country has a natural disaster, the EO satellite operators are tasked with acquiring data, producing maps, and making them available from their websites to support disaster relief activities. During 2021, there were 448 activations of the charter, and while the majority were related to flooding, other activations were for cyclones, volcanoes, earthquakes, fires, and chemical incidents.

10.3.5 Groundwater Measurement

The measurement of groundwater quantifies the amount of water held in the permeable rock beneath the Earth's surface, referred to as the aquifer. It provides 25%–40% of all drinking water worldwide (Morris et al., 2003), and is the primary source of freshwater in many arid countries and one of the conduits by which precipitation falling onto land moves into rivers and lakes. The groundwater level has seasonal variations, which can be adversely affected where water is extracted for activities such as agriculture; excessive extraction leads to questions regarding its sustainability as recharge requires water seeping slowly through soil and rock. When the amount of groundwater falls, it may indicate impending drought, whereas where the amount of groundwater rises, it can cause flooding.

Traditionally, groundwater levels are measured by digging a borehole and putting in an in situ gauge, for either manual or automatic monitoring. Although this indicates the groundwater at that point, it does not provide regional maps. Satellite data can provide an alternative method of monitoring. Techniques include the measurements of deformation (changing shape of the ground) from Interferometric SAR (InSAR) because as the amount of water increases and decreases in the aquifer, it changes the shape of the ground above it.

NASA's Gravity Recovery and Climate Experiment (GRACE) and the follow-on (GRACE-FO) mission, a partnership between NASA and the German Research Centre for Geosciences launched on March 17, 2002, use a different methodology to the passive and active EM radiation discussed previously. Two spacecraft, which fly one behind another, have the changes in their relative distance measured using a K-band microwave ranging system that can be converted to changes in the gravitational field. This approach has enabled GRACE to provide unprecedented insights into mass evolution in the cryosphere and hydrosphere since its launch in 2002; an example of its application to the cryosphere is shown by King et al. (2012), as cited in the Fifth IPCC Report, which has quantified ice mass change over Antarctica. In terms of the hydrosphere, maps of gravity contours can be converted to total water mass, and then when the weight of wet soil is extracted, the amount of aquifer water can be estimated. The NASA JPL GRACE Tellus website (https://grace.jpl.nasa.gov/data/get-data/) provides access to several products, including surface mass density changes as global monthly grids and monthly spatial averages over basins of hydrologic significance.

Weekly maps of groundwater and soil moisture drought indicators derived from GRACE are produced by the United States (US) National Drought Mitigation Center, a collaboration of the University of Nebraska – Lincoln, the Departments of Commerce and Agriculture, and outside experts summarize weekly drought conditions across the United States (Svoboda et al., 2002) (https://droughtmonitor.unl.edu/). Houborg et al. (2012) combined the GRACE data with a long-term meteorological data set within a computer model to generate a continuous record of soil moisture and groundwater dating back to 1948. Figure 10.5w is an example map that shows a drought indicator for the week of March 23, 2015. These maps are aimed at drought associated with climatic variability rather than short-term depletion owing to groundwater withdrawals; if short-term depletion was included, areas such as the southern half of the High Plains aquifer (as discussed next) would always show drought.

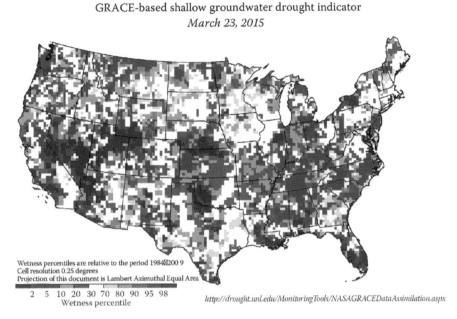

GRACE-based shallow groundwater drought indicator
March 23, 2015

Wetness percentiles are relative to the period 1948–2009
Cell resolution 0.25 degrees
Projection of this document is Lambert Azimuthal Equal Area

2   5   10 20 30 70 80 90 95 98
Wetness percentile

http://drought.unl.edu/MonitoringTools/NASAGRACEDataAssimilation.aspx

FIGURE 10.5
Map of a drought indicator associated with climatic variability for the week of March 23, 2015. (Courtesy of the US National Drought Mitigation Center.)

Other hydrology applications include the work of Moore and Williams (2014), where the surface water storage in lakes was evaluated by GRACE to provide a measure of the total water storage that can be converted to underground water once soil moisture, snow water equivalence, and biomass are accounted for. This technique was also used by Breña-Naranjo et al. (2014) to assess the US High Plains Aquifer, which crosses eight US states from Wyoming to Texas; the aquifer has been used to water crops since the 1950s; however, its underground water level has, on average, fallen by more than 4 m, raising concerns about sustainable water usage. The GRACE data have been used to show this has included a fall of around 0.3 m from 2003 to 2013.

10.4 Practical Exercise: Analysis of the Aswan Dam

For this practical, we will undertake a basic analysis of the area surrounding the Aswan Dam, in Egypt, using microwave SAR data for the first time together with Landsat-8 data and the Shuttle Radar Topography Mission (SRTM)–derived DEM first described in Section 3.2.

The SAR data we will use are an example TerraSAR-X data set provided by Airbus Defence and Space (Airbus DS), which were acquired on June 29, 2013. The collection mode was "Staring SpotLight" with an angle of incidence of 44.5°, a single polarization (HH), and a spatial resolution of 0.36 m.

10.4.1 Step One: Obtaining the TerraSAR-X SAR Data

The TerraSAR-X SAR data needs to be downloaded from the Airbus DS website, and this requires the completion of a very short registration form; as this is an example data set, it's free to use; however, there are some restrictions on the use of the data that can be seen in the license part of the registration.

Go to the Airbus DS website (https://www.intelligence-airbusds.com/), and on the Available Products sidebar, select Radar Imagery > Sample Imagery > Radar Imagery and Data > See all Radar Imagery and Data. Scroll down the screen until you find the Aswan Dam GEC, SE product, and click on it. You'll be taken to the details of the sample imagery, together with a Download option at the bottom. If you click this link, you'll be taken to the short registration form, and after completing it, you'll be sent an e-mail instantaneously with a link to allow a download the 437 MB zipped data file. Once the data have been downloaded and unzipped, you'll have a set of directories containing various SAR data elements.

10.4.2 Step Two: Loading the SAR Data into QGIS

Unzipping will give the folder "2013_06–29_AswanDam_StaringSpotLight_GEC_SE", within this will be the "TSX1_SAR_GEC–SE—ST_S_SRA_20130 629T155330_20130629T155331" folder, and to find the TerraSAR-X data, go into this folder and look for the IMAGEDATA directory, which contains the GeoTIFF (Geostationary Earth Orbit Tagged Image File Format) file of the TerraSAR-X data.

Using the menu item Layer >Add Layer >Add Raster Layer import the "IMAGE_HH_SRA_spot_064.tif" file into QGIS. Once the data have been imported, the screen will be mostly black. Right-click the imported layer, go to Properties > Transparency tab, set the "Additional no data value" to 0 to hide the background values, and press OK.

You'll now likely have a black rectangle with a couple of white spots on the left side, because of the two bright pixels the histogram is too stretched to see the image. Right-click the imported layer, and go the Properties > Histogram and Compute Histogram. The histogram peak is over the far-left side of the plot, and yet the maximum value is over to the far right. Adjust the minimum and maximum values to cover just the peak; we've chosen a range of 10–254. Then, press OK and you should have a view similar to Figure 10.6a.

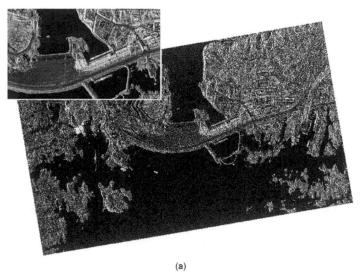

(a)

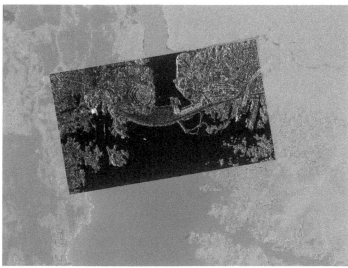

(b)

FIGURE 10.6
Displaying TerraSAR-X data within QGIS as the (a) full image, with a zoomed-in inserted to show the dam area in detail and (b) layers ordered as the TerraSAR-X image, then Landsat-8 NDWI, and finally the Landsat-8 pseudo-true-color composite at the bottom. (Data courtesy of Airbus DS/NASA/USGS.)

You should then see the Aswan Dam in the center of the image, although you'll notice that the image isn't very sharp, which is a trait of SAR imagery and is referred to as being speckled. However, as you zoom in, you'll see the very high spatial resolution of this data set (see Figure 10.6a insert). The noise on the image can be cleaned up by applying one or more of the filters applied in Section 6.12, although this is not something that will be covered in this practical.

The layer has the name IMAGE_HH_SRA_spot_064, and you might find it easier to give it the name TerraSAR-X, which is how it will be referred to in this practical. Right-click the layer name, and there is a Rename Layer option. Rename the layer to "TerraSAR-X".

10.4.3 Step Three: Downloading the Landsat Data from EarthExplorer

The second data layer for this practical will be from Landsat-8, and we'll use EarthExplorer again.

- On the Search Criteria tab, use the Geocoder option and click on the downward-pointing triangle next to Feature (GNIS) to bring up the other options and switch to Path/Row tab, and select the location using WRS-2, with the Path "174" and Row "43". Then press the Show button to move the map to this location.

- The selected location will be highlighted by a blue indicator on the main map, which is slightly to the northeast of Aswan. However, it's important to zoom into the area of interest, the Aswan Dam, until the town of Aswan is at the top of the image in the center, and the majority of Lake Nasser is down the middle of the view.

- Further down the Search Criteria section is the Polygon tab. Press the Use Map button, and you'll see several latitude and longitude coordinates appear in the box above, and the map changes to a shade of red to indicate that the whole area shown has been selected.

- Under the Date Range below, enter 06/15/2013–07/15/2013 as we know the date of the TerraSAR-X image; we ideally want an image close to June 29, 2013. **NOTE:** The dates need to be entered in mm/dd/yyyy format.

- Click on the Data Sets button, or switch to the Data Sets tab at the top. Select Landsat > Landsat Collection 2 Level-1 > Landsat 8–9 OLI/TIRS C2 L1. Finally, press the Results button at the bottom of the screen.

- You'll be offered several pages of Landsat scenes for this area, together with their entity ID, coordinates, acquisition date, path, and row. One of the best ways to look at them is to use the Show

Footprint option, which overlays the Landsat scene onto the map. Turning on and off several images will help you identify the best ones to use. The Aswan Dam is at the top of Lake Nasser, below the town of Aswan.

There is not an ideal image on June 29, 2013, so we have picked the nearest one from July 04, 2013. This image has path 174 and row 43, and the Entity ID LCO8_L1TP_174043_20130704_20200912_02_T1. When you click the Download button, you'll be offered several Landsat products and, like before, it is the Landsat Collection 2 Level-1 Product Bundle you need to download. Click on the Product Options button, and then on the next dialog box, and you'll have the Download option to download the zip file that is 1.16 GB in size.

10.4.4 Step Four: Importing Landsat Data into QGIS

Download and unzip the Landsat data as before. Next, we will import and process it using the Semi-Automatic Classification Plugin (SCP) similar to the Chapter 9 practical. To recap, select the menu item SCP > Preprocessing > Landsat; from the dialog box, choose the directory containing the downloaded Landsat bands you've just unzipped, select the MTL file from that directory, ensure that "Apply the DOS1 atmospheric correction," checkbox is ticked, and if not clicking on it should make the tick appear, press the Run button. You'll need to select the directory where the resulting file is saved to. After a few seconds, the dialog box goes gray, and you'll need to wait for the processing to complete; a bar showing progress is on the main QGIS screen.

Once processing is complete, you'll hear the beeps, and the dialog box will no longer be grayed out. Close the dialog box by pressing X in the top right corner. Each of the individual spectral wavebands will be loaded in QGIS and listed in the Layers panel, and the image will be black and white.

10.4.5 Step Five: Creating an NDWI Using a Mathematical Function

The next stage involves using a mathematical function to create NDWI, separating land from water, as described in Equation 10.1, using the green and NIR wavebands that are 3 and 5 for the Landsat-8 OLI data, respectively.

Go to the menu item SCP > Band calc, and press the Refresh list button on the dialog box that appears to ensure that all the Landsat wavebands are recognized. You should see the list of wavebands and beneath it the expression area, with mathematical functions to the right. Creating an equation is simply a case of single clicks on the mathematical functions and double clicks on the wavebands. For the NDWI, the formula in the expression band should be

$$\frac{("RT\_LCO8\_L1TP\_174043\_20130704\_20200912\_02\_T1\_B3" - "RT\_LCO8\_L1TP\_174043\_20130704\_20200912\_02\_T1\_B5")}{("RT\_LCO8\_L1TP\_174043\_20130704\_20200912\_02\_T1\_B3" + "RT\_LCO8\_L1TP\_174043\_20130704\_20200912\_02\_T1\_B5")} \quad (10.3)$$

Press the Run button, and you'll be prompted for a location and name to save the output GeoTIFF file; call it NWDI. Once the Save is pressed, the processing will start with the dialog box going gray; the processing time should be shorter for this step. After beeps, close the dialog box; you'll see that there is an NWDI layer.

Right-click on this layer, and select Properties, go to the Symbology tab at the top and change the Render type to "Singleband pseudocolor". Then use the Color ramp option further down to apply a color map; we've used Blues as this will be a water layer. Press OK. The image will now have a range of blues, with the colors and numbers representing the NDWI ratio values, with dark colors/higher numbers (all negative in this practical) representing pixels that are more likely to be water, which, in our case, should correspond to Lake Nasser and the river.

Finally, in terms of layer order, put the TerraSAR-X layer on the top followed by the NDWI, and turn off all the individual wavebands. If you zoom out, you should have the NDWI, with the SAR image of the Aswan Dam on top.

The NDWI ratio is used to separate land from water and, in this practical, the land and the water are easy to visually separate. However, in flood mapping, where it's less evident, this approach can be a powerful tool.

10.4.6 Step Six: Creating a Pseudo-True-Color Composite

This step will create an RGB pseudo-true-color composite from the Landsat data, to allow a comparison between the derived NDWI data and the surrounding terrain. For this step, select menu item Processing > Toolbox, and you'll open the Toolbox panel within the right sidebar. In the Toolbox, using the right-pointing triangle, select GDAL > Raster Miscellaneous and then double-click on Merge.

For the input layers, press the ... button and select wavebands 4, 3, and 2 by using the checkboxes, and put them to the top in that order. Click OK to return to the Parameters tab, click the check box for "Place each input file into a separate band", then press Run.

Once the processing has been completed, you'll see a message at the bottom of the log "Algorithm 'merge' finished". Click Close. A new layer will appear in the layer panel, named Merged; we'd recommend renaming this to "Pseudo-True Color", so you know what it is.

Right-click on the layer, and select Properties. Go to the Histogram tab, Compute Histogram, and then change the minimum and maximum values for each band; we've chosen the red as 500–1,300, green as 800–2,100, and blue as 1,200–3,300. Go to the Transparency tab of Properties, put 0 in the "Additional no data value" box, then change the Global Opacity value to 50% using the slider or entering the values manually.

Finally, you need to adjust some properties on the NDWI layer. Right-click the NDWI layer, click Properties, and select the Symbology tab. You should see a table with Value, Color, and Label representing this layer's different shades of blue. Double-click on the top color in the list, and a "Select color" dialog box will open. On the right side, toward the bottom of the list, there is a setting for Opacity. Change the opacity percentage from one hundred to zero, and press OK. On the color list, this color will appear as a white and gray checkboard, which means it is set to be transparent. Repeat this with every color in the color list that has a negative figure in Value. You're effectively making the land transparent and keeping the water pixels.

If you ensure the layer order is TerraSAR-X, NDWI, and Pseudo True Color, with all other layers turned off. You should now see the Aswan Dam in black and water; underneath Lake Nasser water will be blue, and the land will be in varying colors; as shown in Figure 10.6b.

10.4.7 Step Seven: Downloading the SRTM DEM Data

The next step is to download the SRTM DEM data. This can be done from several websites; however, we will continue using EarthExplorer.

To get the SRTM DEM data, go to the EarthExplorer website and repeat what we did earlier for getting the Landsat data; to recap, enter the WRS path and row number for the Landsat scene, zoom in to select the majority of Lake Nasser with Aswan at the top center of the image, and press Use Map.

If you're working through the practical from start to finish, you'll still have June to July 2013 in the date range, which won't work for the STEM DEM data. Therefore, go into each of the Search From and Search To dates, and press your Delete key; this will reset the dates.

Choose the Data Sets tab, turn off the Landsat data if it is still selected, and choose instead Digital Elevation > SRTM > SRTM Void Filled, and then press the Results button. Using the Void Filled data set means that any areas of missing data, termed voids, are filled using interpolation algorithms in conjunction with other sources of elevation data.

The result will be a series of SRTM data tiles (effectively an image) to download, together with the Entity ID (unique reference number), publication date (01-OCT-12), resolution (ARC-3), and coordinates. For this practical, you'll need four tiles with the following coordinates:

- 23, 32
- 23, 33
- 24, 32
- 24, 33

NOTE: You may have to scroll down to find them, or they may even be on a subsequent results page. To download a tile, select the Download button. You'll be given several options and will need the "GeoTIFF 3 Arc-second" product; download this file for each tile.

10.4.8 Step Eight: Loading the SRTM DEM Data into QGIS

You don't need to unzip these files as they're already GeoTIFFs and can be directly loaded into QGIS using the menu item Layer > Add Layer > Add Raster Layer; multiple files can be uploaded at the same, using the CTRL key. The four files to load are as follows:

- n23_e032_3arc_v2.tif
- n23_e033_3arc_v2.tif
- n24_e032_3arc_v2.tif
- n24_e033_3arc_v2.tif

When the four SRTM DEM tiles have been loaded into QGIS, you'll have a much larger area of Egypt than the other images; if you turn the other layers off and keep the SRTM DEM layers, then zoom out, you'll see all four tiles. The edges of the tiles will be visible as QGIS automatically stretches each tile slightly differently.

NOTE: If only a black screen is visible, it probably means the view is zoomed in too much. Therefore, zoom out, and the four tiles will become visible.

10.4.9 Step Nine: Merging the Four SRTM DEM Tiles into a Single Layer

From the toolbox panel on the right sidebar, select GDAL > Raster Miscellaneous, and then double-click on the Merge raster layers option. A dialog box will appear, and for the Input layers, press the ... button, select the four SRTM layers, click OK and then click Run.

This merging will create a new layer called "Merged" that looks like the four SRTM tiles displayed together, but the edges have disappeared as all the data have a single histogram stretch applied. Rename the layer to "SRTM DEM".

To enhance the features, right-click the SRTM DEM layer, and select Properties. On the Symbology tab:

- Change the Render type to single-band pseudocolor.
- Change the minimum and maximum values – we've chosen 77–550.
- Press the downward-pointing triangle at the end of Color ramp and select a new color scheme for your image – we've chosen the Spectral from the drop-down list. The colors, and associated range of numbers, represent the altitude (height) of the land above sea level in meters. The colors, from blue to red, are used to show the altitude of the terrain where blue is low heights and blue is higher.
- Press the downward-pointing triangle at the end of Color ramp again, and this time select the Invert Color Ramp option. This will mean that the lower figures start in blue and go to red for the higher ones, which we think is more intuitive.
- Press OK.

Turn off all the layers except the SRTM DEM layer. Then order the layers with TerraSAR-X on top, followed by the NWDI, Pseudo-True Color composite, and finally SRTM DEM; you will see a multi-colored image similar to that shown in Figure 10.7a.

10.4.10 Step Ten: Adding Contour Lines

It's possible to create contour lines from the SRTM DEM, which can be overlaid onto the raster products to indicate height. Go to the menu item Raster > Extraction > Contour.

For the input file, the SRTM DEM layer should be automatically populated, but select it from the drop-down menu if not, and change the Interval between contour lines to 100.00000 m so you don't get too many contour lines. Press Run; when the processing is completed, press Close. This processing will have added a Contour layer to the Layers panel. Using the Symbology tab, the contours' color, thickness, and style can be changed if required.

Finally, turn on the TerraSAR-X, NDWI, and pseudo-true-color layers, and turn off all other layers. Reorder the layers, so the TerraSAR-X image is on top, followed by the contours, NDWI, and finally, the pseudo-true-color composite. This final image (Figure 10.7b) shows information about where the open water is and the terrain surrounding the Aswan Dam and Lake Nasser.

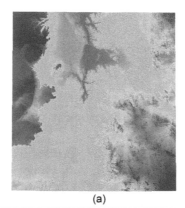

(a)

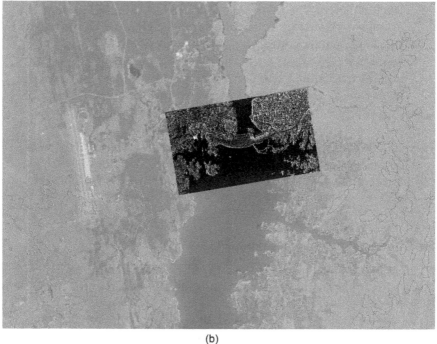

(b)

FIGURE 10.7
(a) Shuttle Radar Topography Mission (SRTM) tiles with TerraSAR-X image overlaid and (b) layers ordered as the TerraSAR-X image, then SRTM DEM derived 100 m contours, then Landsat-8 NDWI, and finally the Landsat-8 pseudo-true-color composite at the bottom. (Data courtesy of Airbus DS/NASA/USGS.)

10.5 Summary

This chapter focused on inland water bodies and the terrestrial water cycle. The theory element provided a detailed review of microwave data for the first time, which continued into the practical with SAR imagery. We've also introduced DEM data, commonly included within hydrological modeling, to add terrain elevation and contours to the imagery.

As the applications in this chapter are directly related to aspects of climate change, the skills learned offer opportunities for you to do your research into different elements of how the planet's climate is evolving.

10.6 Online Resources

- Airbus DS sample imagery: https://www.intelligence-airbusds.com/
- AVISO website containing altimetry data for marine and hydrological applications: https://www.aviso.altimetry.fr/en/home.html
- CoastWatch Great Lakes program: https://coastwatch.noaa.gov/cw/index.html
- Copernicus Global Land Service water-level products: https://land.copernicus.eu/global/products/wl
- EarthExplorer browser: https://earthexplorer.usgs.gov/
- ESA funded Soil Moisture Climate Change Initiative (CCI) project: https://climate.esa.int/en/projects/soil-moisture/
- ESA Climate Change Initiative (CCI) data website: https://climate.esa.int/en/
- HYDROLARE project http://hydrolare.net/
- International Charter Space and Major Disasters: https://www.disasterscharter.org/
- NASA GPM missions' website: https://gpm.nasa.gov/
- NASA JPL GRACE Tellus website: https://grace.jpl.nasa.gov/data/get-data/
- US Department of Agriculture's Foreign Agricultural Service Crop Explorer portal: https://ipad.fas.usda.gov/cropexplorer/Default.aspx

10.7 Key Terms

- Aquifer: The amount of water held in the permeable rock beneath the Earth's surface.

- Contour: A line joining parts of a landscape with the same height.
- Digital elevation model: A raster or vector point layer that contains height values.
- Normalized difference water index: Separates water from other pixels within an image.
- Terrestrial water cycle: The cycle of rain, evaporation, and restocking of the groundwater, rivers, and lakes.

References

Berry, P. A. M., J. D. Garlick, J. A. Freeman and E. L. Mathers. 2005. Global inland water monitoring from multi-mission altimetry. *Geophys Res Lett* 32:L16401. Available at http://doi.org/10.1029/2005GL022814.

Birkett, C. M. and B. Beckley. 2010. Investigating the performance of the Jason-2/OSTM radar altimeter over lakes and reservoirs. *Mar Geod* 33:204–238.

Breña-Naranjo, J. A., A. D. Kendall and D. W. Hyndman. 2014. Improved methods for satellite-based groundwater storage estimates: A decade of monitoring the high plains aquifer from space and ground observations. *Geophys Res Lett* 41:6167–6173.

Crétaux, J.-F., W. Jelinski, S. Calmant et al. 2011. SOLS: A Lake database to monitor in near real time water level and storage variations from remote sensing data. *Adv Space Res* 47(9):1497–1507.

FAO. 2015. World food summit: Factsheet water and food security, AD/I/Y1300E/1/7.01/36000. Available at http://www.fao.org/worldfoodsummit/english/fsheets/water.pdf (accessed April 17, 2015).

Folland, C. K., T. R. Karl, J. R. Christy et al. 2001. Observed climate variability and change. In *Climate Change 2001: The Scientific Basis*, eds. J. T. Houghton, Y. Ding, D. J. Griggs, M. Noguer, P. J. van der Linden, X. Dai, K. Maskell and C. A. Johnson, 1–83. Cambridge: Cambridge University Press.

Frappart, F., S. Calmant, M. Cauhope, F. Seyler and A. Cazenave. 2006. Preliminary results of ENVISAT RA-2-derived water levels validation over the Amazon basin. *Remote Sens Environ* 100:252–264.

Houborg, R., M. Rodell, B. Li, R. Reichle and B. F. Zaitchik. 2012. Drought indicators based on model-assimilated Gravity Recovery and Climate Experiment (GRACE) terrestrial water storage observations. *Water Resour Res* 48:W07525. Available at http://doi.org/10.1029/2011WR011291.

King, M. A., R. J. Bingham, P. Moore, P. L. Whitehouse, M. J. Bentley and G. A. Milne. 2012. Lower satellite-gravimetry estimates of Antarctic sea-level contribution. *Nature* 491(7425):586–589.

Lavender, S., P. Healy, I. Robinson et al. 2017. Application of earth observation to a Ugandan drought and flood mitigation service. In *2017 Big Data from Space Conference (BiDS'17)*, Toulouse, 28–30 November 2017: 469–472.

McFeeters, S. K. 1996. The use of normalized difference water index (NDWI) in the delineation of open water features. *Int J Remote Sens* 17:1425–1432.

Moore, P. and S. D. P. Williams. 2014. Integration of altimetric lake levels and GRACE gravimetry over Africa: Inferences for terrestrial water storage change 2003–2011. *Water Resour Res* 50:9696–9720.

Moreira, A., G. Krieger, I. Hajnsek et al. 2015. Tandem-L: A highly innovative bistatic SAR mission for global observation of dynamic processes on the Earth's surface. IEEE Geoscience and Remote Sensing Magazine 3(2):8–23. http://doi.org/10.1109/MGRS.2015.2437353.

Morris, B. L., A. R. L. Lawrence, P. J. C. Chilton et al. 2003. Groundwater and its susceptibility to degradation: A global assessment of the problem and options for management. Early Warning and Assessment Report Series, RS. 03-3. United Nations Environment Programme, Nairobi, Kenya. Available at http://www.unep.org/dewa/water/groundwater/pdfs/Groundwater_INC_cover.pdf (accessed April 17, 2015).

Palmer, S. C. J., T. Kutser and P. D. Hunter. 2015. Remote sensing of inland waters: Challenges, progress and future directions. *Remote Sens Environ* 157:1–8.

Svoboda, M., D. Lecomte, M. Hayes et al. 2002. The drought monitor. *Bull Am Meteorol Soc*, 83(8):1181–1190. http://doi.org/10.1175/1520-0477(2002)083.

Vörösmarty, C. J. 2009. Chapter 10, The Earth's natural water cycles. United Nations World Water Development Report 3rd Edition. Available at http://webworld.unesco.org/water/wwap/wwdr/wwdr3/pdf/22_WWDR3_ch_10.pdf (accessed April 17, 2015).

Wolfgang Kron, I. 2005. Storm surges, river floods, flash floods—Losses and prevention strategies. Risk factor of water Special feature issue, Touch Publication, Munich Resinsurance. Available at https://www.munichre.com/site/touch-publications/get/documents_E832695857/mr/assetpool.shared/Documents/5_Touch/_Publications/302-04693_en.pdf (accessed April 17, 2015).

Xu, H. 2006. Modification of the normalised difference water index (NDWI) to enhance open water features in remotely sensed imagery. *Int J Remote Sens* 27(14):3025–3033.

11

Coastal Waters and Coastline Evolution

Coastal zones are places where the sea and the land meet, and they have social, economic, and environmental importance as they attract human settlements and economic activity. The habitats include beaches, sand dunes, marshes, river mouths, coral reefs, and mangroves. These support recognizable ecosystem services, including tourism, food production, and wildlife, and less obvious ones, such as carbon and pollution sinks. An illustration of the importance of these areas comes from a review by Gedan et al. (2011), which concluded that coastal wetland plants interact with water and sediment in a variety of direct and indirect ways to slow water flow and facilitate sediment deposition, increasing shoreline cohesion and providing natural protection.

However, coastal zone habitats are often under threat from urban developments, floodplains, coastal reclamation, and changes in agriculture and food production; in the longer term, climate change impacts. Remote sensing is used to research and monitor how coastal zones are evolving and help identify why.

11.1 Optical Data

11.1.1 The Color of the Water

The color of water bodies, termed *ocean color*, was recorded as early as the 1600s when Henry Hudson noted in his ship's log that a sea pestered with ice had a black–blue color (Hudson, 1868). This color relates to both the sea surface–reflected radiance and the spectral variation in the water-leaving radiance $L_w(\lambda)$, that is, the radiance signal leaving the water after the solar irradiance has interacted with what's in the water. By measuring the spectral variations in the water-leaving radiance, it's possible to extrapolate what type of (organic and inorganic) material is dissolved and suspended in the water column. Although the individual particles in the water are far smaller than the spatial resolution of satellite sensors, they become detectable and measurable with a sufficient concentration.

To measure ocean color, the first step is to convert water-leaving radiance to remote sensing reflectance, $R_{rs}(\lambda)$ (sr^{-1}), by dividing it by the downwelling irradiance at the sea surface, as shown in Equation 11.1. This correction removes the variations in the solar irradiance and gives a more accurate representation of both the spectral and magnitude variations:

$$R_{rs}(\lambda, z) = \frac{L_w(\lambda, z)}{E_d(\lambda, z)}. \qquad \text{11.1)}$$

It's also very common, when measuring ocean color from satellites, to apply an atmospheric correction (AC) to remove the atmospheric and sea surface signals, transforming reflectance from top-of-atmosphere (TOA) values to bottom-of-atmosphere (BOA) values, essentially calculating $R_{rs}(\lambda)$. It's important as only 5%–10% of the TOA signal originates from the water, whereas for the land, it's 10%–50%. However, as noted in Chapter 10, applying an AC is becoming increasingly common in all remote sensing applications because of the inconsistency of algorithm results without this correction.

Once ocean color data have been atmospherically corrected, they can be used to derive quantitative parameters, such as the concentration of the organic and inorganic materials. The material suspended in water is known as suspended particulate matter (SPM), which can be divided into two main groups:

- Biological particles dominated by phytoplankton, minute micro-algae creatures, which primarily contain the photosynthetic pigment chlorophyll-a (Chlor-a) and account for approximately 50% of the photosynthesis on Earth, and
- Inorganic particles such as the sediment brought up from the seabed, as well as the skeletons of phytoplankton, such as diatoms with a silicon shell and coccolithophore that are covered in calcium carbonate plates.

In addition, there's dissolved material in the water resulting from the land surface and subsurface runoff or biological degradation; referred to as colored dissolved organic matter (CDOM).

The effect of these materials on the water-leaving radiance depends on their absorption and scattering properties, which vary nonlinearly in terms of both wavelength and changes in concentration. Therefore, extracting accurate ocean color information from imagery requires the careful application of algorithms and an understanding of optical theory.

The simplest approach is correlating a single spectral waveband with the material of interest and determining a statistical relationship.

For example, the natural logarithm of SPM and the reflectance in the red wavelength region are often strongly correlated. An alternative approach is to use waveband ratios as they suppress solar angle and atmospheric effects and may also cancel out effects caused by the sensor pixel geometry.

Focusing on phytoplankton, it's not the phytoplankton concentration that's primarily measured; instead, it's the level of chlorophyll-like pigments in the phytoplankton that interacts with the light, referred to as the pigment concentration or simplified to the Chlor-a concentration.

Phytoplankton prefers nutrient-rich waters, and thus the concentrations vary from less than 0.03 mg/m³ in waters poor in nutrients up to more than 30 mg/m³ in nutrient-rich waters; when there is a very high concentration of phytoplankton in an area, it's often described as a phytoplankton bloom. As the phytoplankton concentration increases, the reflectance in the blue waveband decreases, owing to the absorption of chlorophyll-like pigments, while the reflectance in the green waveband increases slightly because of scattering. Therefore, a ratio of blue to green remote sensing reflectance can be used to derive quantitative estimates of the pigment concentration; the first quantitative work was by Clarke et al. (1970) using an airborne spectroradiometer.

Similar to when Normalized Difference Vegetation Index as was discussed in Chapter 9, different sensors operate at different wavelengths and, hence, care needs to be taken when choosing the product to use for assessing ocean color. Standard Chlor-a products measure phytoplankton pigment concentrations using a blue/green band ratio formulated initially at the SeaWiFS Bio-optical Algorithm Mini-workshop called the Ocean Color 2 (OC2) algorithm (O'Reilly et al., 1998). At the time of writing, the algorithm is called OC4; the names and wavebands vary slightly in algorithms between the different ocean color missions because of their sensor differences and are developing over time (Hu et al., 2012). At high pigment concentrations, the signal at 440 nm becomes too small to be retrieved accurately. Thus, the pigment algorithm switches from a blue/green to a green/red ratio that's less sensitive to variations in the pigment concentration.

Examples of other algorithms used to monitor phytoplankton include the following:

- Chlorophyll fluorescence: Measuring the fluorescence of the pigments through the remote sensing reflectance can also be used to determine the amount of pigment concentration. It's useful as it is closely linked to cell physiology and can vary with nutrient status, species composition, and growth rate (Gower and Borstadt, 1990). Medium Resolution Imaging Spectrometer (MERIS) and MODIS have a relatively narrow waveband at 682.5 nm, which can detect the fluorescence signal (Rast et al., 1999).

- Maximum Chlorophyll Index (MCI) supplements chlorophyll fluorescence, and it's traditionally applied to the TOA reflectance. It's used in coastal waters where there can be high concentrations of inorganic sediments as well as phytoplankton.

In clear oceanic waters, the reflectance is primarily a function of the scattering and absorption of the water and algal pigments, and Chlor-a concentration can be determined accurately. In coastal waters, the analysis becomes more complicated because of the scattering and absorption from the inorganic sediment particles, as they strongly backscatter radiation that overwhelms the reflectance signal of phytoplankton pigments.

Figure 11.1 shows different versions of an ocean color image using MODIS-Aqua data acquired on September 8, 2002. Figure 11.1a shows the full scene as the TOA pseudo-true-color composite reflectance image (with wavebands 1, 4, and 3 shown as red, green, and blue), with the New York Bight area in the top left. However, the detail is difficult to see because the clouds and land have much higher reflectance than the water. Figure 11.1b is then a zoomed-in version of the BOA pseudo-true-color reflectance image (using the R_{rs} 667, 431, and 443 nm wavebands), that is, after AC. Finally, Figure 11.1c is the Chlor-a map created using the OC3M algorithm. Offshore, the Chlor-a concentrations are low (less than 5 mg/m³) and are believable. However, care should be taken when interpreting higher Chlor-a values (greater than 5 mg/m³ close to the coastline) as it could be phytoplankton blooming, but it could also be high concentrations of sediment adversely affecting the algorithm. Values should also be investigated using processing flags and the pseudo-true-color composite.

11.1.2 Bathymetric Data

If the water body is sufficiently shallow, then both active and passive remotely sensed data can also be used for mapping water depth, a technique commonly called bathymetry.

The formative approach to deriving bathymetric data defined by Lyzenga (1978) assumed an approximately linear function between bottom reflectance and an exponential function of the water depth. More complex techniques exploit the fact that different spectral wavelengths are attenuated by water to differing degrees, with red light being attenuated much more rapidly than blue wavelengths. Like altimetry, the depth from Lidar is calculated from the time the pulse takes to leave the emitter, be reflected off the bottom, and be picked up by the receiver. The attenuation at the operating wavelength, normally green, needs to be sufficiently low that the pulse is not fully absorbed by the water, and there's a near-infrared wavelength to determine the water surface.

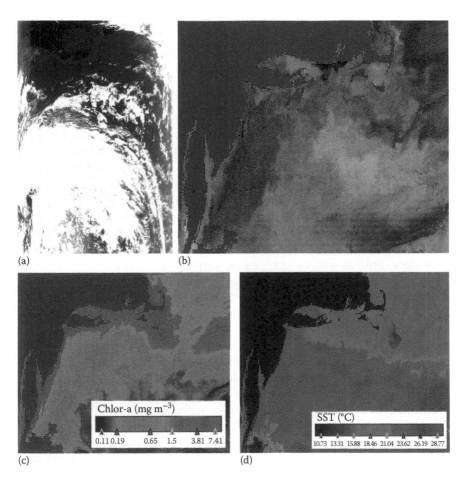

FIGURE 11.1

MODIS-Aqua imagery for September 8, 2002, shown as the (a) pseudo-true-color composite full scene TOA reflectance image alongside the zoomed-in image to show the New York Bight area as the (b) pseudo-true-color composite BOA reflectance image, (c) Chlor-a map, and (d) SST. (Data courtesy of NASA.)

Both passive and active data sources are negatively influenced by the reflection of solar radiation from mirror-like water surfaces, termed *sun glint*. However, Lidar data can be operated at night to avoid this error source.

Bathymetric Lidar measurements have, to date, only been operated from aircraft. The use of passive satellite data has increased with the availability of affordable, but not free, high spatial resolution imagery such as WorldView-2 and WorldView-3, as these sensors have been optimized with a so-called coastal waveband from 400 to 450 nm; see Section 3.1 for more details.

11.2 Passive Microwave Signatures from the Ocean

The theory for active microwave signals from the ocean is based on altimetry and Synthetic Aperture Radar (SAR), which were discussed in Chapter 10; hence, in this section, we're going to focus on passive microwave signatures used for measuring sea surface temperature (SST), salinity, and sea ice.

SST measures the water temperature close to the ocean's surface. The ocean's emissivity is relatively constant for the SST measurement frequencies, and thus variations in the brightness temperature can be directly related to changes in the physical water temperature; brightness temperature was defined in Chapter 8, with the conversion from a microwave radiance to temperature following the same process for the ocean's signal. Therefore, for an idealized ocean, it would be possible to infer SST from brightness temperature alone. Still, as with optical remote sensing, there is a need to account for the nonideal surface, such as the roughening effects from the wind. Microwave SST data collection tends to have large footprints, owing to the low signal level. Hence, care needs to be taken around coastal waters where the footprint includes both the relatively bright land and darker water.

Microwave and thermal SST measurements have their own strengths and weaknesses, with the accuracy and resolution of microwave measurements being poorer than that of thermal infrared (TIR) measurements. Still they are unaffected by clouds and generally easier to correct for atmospheric effects.

Salinity in the ocean is measured as the number of grams of salt per 1,000 g of water and, like SST, is measured at the ocean's surface. The technique is very similar to SST because it can be derived from brightness temperature, although salinity only influences emissivity at short frequencies of less than 6 GHz. It's a much newer technological approach than SST, with the Soil Moisture and Ocean Salinity (SMOS) Microwave Imaging Radiometer with Aperture Synthesis (MIRAS) instrument being a two-dimensional interferometric passive microwave sensor, as initially explained in Chapter 3. The Aquarius sensor followed it in 2011, which was carried by the Argentine SAC-D satellite, which is a combined 1.4-GHz polarimetric radiometer and 1.26-GHz Radar, on a mission also carrying a microwave radiometer and TIR sensor allowing the capture of SST through both thermal and microwave sensors, with a radar scatterometer to measure, and correct for, the effects caused by ocean waves. Aquarius was the first mission with a primary focus of measuring sea surface salinity. Finally, it's also worth noting that sensors can also pick up stray noise from other sources (Hallikainen et al., 2010); for example, mobile phones operate in the L band (1.4 GHz), which is the same frequency that SMOS and Aquarius operate at.

Passive microwave sensors on satellites have also been used to measure sea ice thickness from the 1960s onward as they provide good spatial and temporal coverage alongside being almost independent of cloud coverage and daylight conditions. As reflected microwave energy depends on the target's physical properties, the distinction between sea ice and ocean is straightforward.

Starting in 1987, the Special Sensor Microwave Imager provided continuous measurements of sea ice thickness as part of the US Defense Meteorological Satellite Program for around 30 years. In 2015, the US Congress voted to terminate the DMSP program and to scrap the last unlaunched DMSP 5D-3/F20 satellite, ordering the Air Force to move on to a next-generation system. In addition, DMSP-5D-3/F19 stopped responding to commands in February 2016 and has not provided full-orbit weather imagery since. In addition, the Advanced Microwave Scanning Radiometer-Earth Observing System (AMSR-E) sensor onboard NASA's Aqua satellite has provided derived products such as precipitation rate, SST, sea ice concentration, snow water equivalent, soil moisture, surface wetness, wind speed, atmospheric cloud water, and water vapor.

SAR and optical data can be used to provide detailed images of sea ice, which is used to help route ships through ice-covered regions; the RADARSAT satellites managed by the Canadian Space Agency are focused on this activity, and the Sentinel-1 satellites of the European Space Agency (ESA) is now providing freely available public data.

11.3 Coastal Applications

11.3.1 Physical Oceanography that Includes Temperature, Salinity, and Sea Ice

Physical oceanography covers the physical properties of the ocean (including temperature, salinity, and sea ice) and how these elements contribute to the exchange of heat and moisture between the ocean and the atmosphere. These applications are mainly based on passive microwave remote sensing, as discussed in Section 11.2.

SST is an important geophysical parameter that helps define the boundary of the air–sea interface, which is used in estimating heat energy transfer, known as the heat flux. It also contributes insights into global atmospheric and oceanic circulation patterns together with anomalies such as the El Niño Southern Oscillation, associated with a band of warm ocean water that develops in the central and east-central Pacific Ocean. At the same time, on a local scale, it traces eddies, fronts, and upwelling areas that are important for marine navigation and biological productivity.

SST is also valuable for commercial fisheries exploitation, using the established relationships between fish schools and thermal fronts (Simpson, 1994). Global data sets are available from the Group for High-Resolution Sea Surface Temperature (GHRSST, https://www.ghrsst.org/).

Sea surface salinity is directly linked to the global water cycle that governs water exchange between the land, oceans, and atmosphere. High- and low-salinity patches can be related to evaporation and precipitation, respectively, and it's also a key variable in ocean circulation because it influences the water density and density-driven currents. As mentioned in Section 11.2, sea surface salinity is relatively new, and the primary data sets are available from SMOS and Aquarius. However, it has also been determined optically in coastal environments when there can be a strong relationship between salinity and CDOM.

Sea ice acts as a thermally insulating layer reflecting electromagnetic (EM) energy, keeping the atmosphere above it cold. As described in Section 11.2, the AMSR-E sensor onboard Aqua provides a variety of products that can be downloaded through the US National Snow and Ice Data Center (NSIDC, https://nsidc.org/data/amsre/index.html); another source of data is the EUMETSAT Ocean and Sea Ice Satellite Application Facility (OSI SAF, https://osi-saf.eumetsat.int/products/sea-ice-products/). For example, Figure 11.2 is an AMSR-E L3 gridded product (AE_SI12) showing sea ice concentration percentages for Antarctica in October 2002 on a 12.5-km spatial resolution using a polar stereograph grid projection.

SMOS can also be used to measure sea ice thickness. In contrast to the higher-frequency microwave missions, it has a signal that originates not only from the ice surface but also from deeper down and from the water underneath the ice if the thickness is less than 1 m (Tian-Kunze et al., 2014). Satellites focused on measuring sea ice thickness include NASA's Lidar-based Ice, Cloud and Land Elevation Satellite-Geoscience Laser Altimeter System (ICESat-GLAS), which operated between 2003 and 2010, followed by ICESat-2 in September 2018, and the ESA CryoSat-2 satellite launched in 2010, introduced in Chapter 10.

11.3.2 Water Quality, Including Algal Blooms

This application primarily uses ocean color measurements alongside SST and sometimes SAR/optical imagery for the movement of water through ocean currents. For ocean color, water quality can mainly only be detected and mapped if it influences the water-leaving radiance; exceptions can be parameters such as nutrients where an indirect relationship can be established with optically active materials or oil that influences the sea surface reflectance.

Coastal waters are productive and sensitive marine ecosystems. Chlor-a is an important water quality parameter as phytoplankton provide food for other aquatic life but may also indicate pollution sources such as high

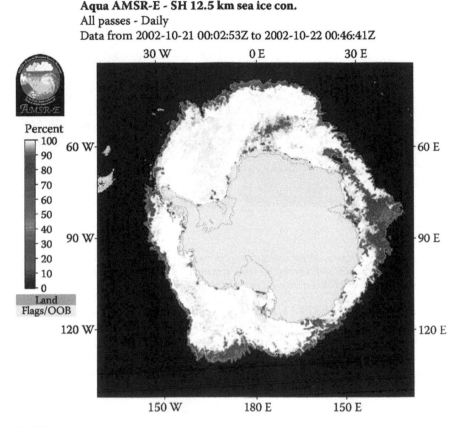

Aqua AMSR-E - SH 12.5 km sea ice con.
All passes - Daily
Data from 2002-10-21 00:02:53Z to 2002-10-22 00:46:41Z

FIGURE 11.2
AMSR-E sea ice data browse image for Antarctica February 28, 2014. (Data courtesy of
Cavalieri et al. (2004), NSIDC; Copyright © 2014 The University of Alabama in Huntsville.
All rights reserved.)

nutrient inputs from land drainage. Harmful algal blooms (commonly
shortened to HABs or called red or brown tides on account of the red
or brown coloration given to the water by high concentrations of specific
phytoplankton species) can have toxic effects. Shellfish filter large quantities of water and concentrate the algae in their tissues, which can directly
affect the aquaculture industry by mass fish mortalities and indirectly as
shellfish accumulate toxins that can, under certain circumstances, render
them unsafe for human consumption.

Blooms occurring in coastal waters are more challenging to detect
than those in the open ocean as the standard Chlor-a algorithm may pre-
dict high levels of Chlor-a owing to SPM reflectance or CDOM absorption rather than Chlor-a absorption of a bloom, as first mentioned in
Section 11.1. Figure 11.3 shows phytoplankton blooming off the coast of

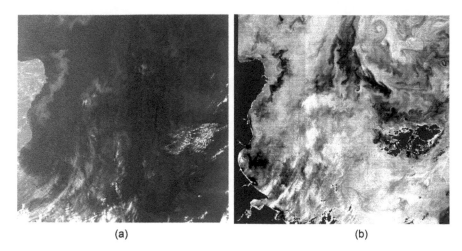

(a)   (b)

FIGURE 11.3
Phytoplankton blooming off the coast of Argentina using MERIS Level 1 reduced resolution imagery captured on February 10, 2003, shown as the (a) TOA radiance pseudo-true-color composite and (b) MCI image. (Data courtesy of ESA/NASA.)

Argentina using MERIS L1 Reduced Resolution (1-km pixels) acquired on February 10, 2003. Figure 11.3a shows the ocean color signal as the TOA radiance pseudo-true-color composite, while Figure 11.3b shows MCI calculated using the module available in the Sentinel Application Platform (SNAP). The TOA image is dominated by the cloud and the land with much higher reflectances than the water. However, highly reflective phytoplankton blooms (likely to be coccolithophores) are visible alongside the Argentinian coastline on the left. When the MCI is applied, the in-water features become much more visible, and the land is now a homogenous gray owing to a mask. Coccolithophore blooms have low MCI values as they're covered in calcium carbonate plates and thus primarily act as scatters rather than having a strong signal from the Chlor-a absorption.

Monitoring a bloom once it has been detected is important, but it would also be desirable to be able to forecast the potential occurrence of blooms to support industries such as aquaculture. Therefore, studying the local environment will provide information on the factors that influence the initiation of a bloom and its evolution, and spectral information can be used to identify different species' characteristic absorption and scattering properties together with their abundance. This information can provide an assessment of the potential risks of blooms occurring.

Landsat-8 and Landsat-9, together with the Copernicus Sentinel-2 mission, improve opportunities for ocean color processing due to increases in wavebands and spectral resolution that will aid the remote sensing of the nearshore coastal environment, including estuaries.

11.3.3 Mangroves and Coastal Protection

Mangrove and wetland forests are among the most important ecosystems from biodiversity, carbon sequestration, and economical standpoints. Mangroves occur along oceanic coastlines throughout the tropics and subtropics, between the latitudes of 38°S and 31°N, with geographical regions including Africa, the Americas, Asia, and Australia. They support numerous ecosystem services, including nutrient cycling and fisheries, being highly productive with a rich diversity of flora and fauna. However, in 2007, it was reported that the areal extent of mangrove forests had declined by between 35% and 86% over the last quarter century as a result of coastal development, aquaculture expansion, and overharvesting (Duke et al., 2007) with the disturbance ranging from pollution to direct clearance (Alongi, 2002). This loss of mangrove forests significantly reduces a shoreline's capacity to resist storm surges; for example, Fritz and Blount (2007) noted that, in Bangladesh, 600 m of mangroves and trees reduced storm surges by 7 m.

Remote sensing techniques have been used extensively to study mangroves, with the Landsat and Sentinel-2 satellites being commonly used alongside high-resolution imagery from IKONOS and QuickBird, providing a baseline mangrove database for their future monitoring. Vegetation indices are applied to discriminate between areas of high versus low biomass and healthy versus unhealthy status. Green et al. (1998) applied five different methodologies to Landsat Thematic Mapper (TM), SPOT XS, and airborne imagery over the Turks and Caicos in the eastern Caribbean and concluded that an accurate classification was possible from both the Landsat and airborne imagery. The most accurate Landsat method was the use of a supervised maximum likelihood classification with Principal Components Analysis (PCA) and band ratio products as inputs; maximum likelihood is another option within the "Semi-Automatic Classification" Quantum Geographic Information System (QGIS) plugin described in Chapter 9.

The Cayman Islands are positioned within the Caribbean Hurricane Belt, and woody vegetation, such as mangroves, provide coastal protection as recognized by the Cayman Island National Trust. Figure 11.4 shows the PCA unsupervised classification of July 16, 2009, SPOT image of Grand Cayman, where Figure 11.4a is the first, Figure 11.4b is the second, and Figure 11.4c is the third principal component. The first three principal components, as discussed in Section 9.2, show the majority of the information with the first component giving a strong separation between land and water. Then, the second and third components, respectively, enhance features in the urban areas at the western end of the island and vegetation features at the eastern end of the island.

SAR interferometry also allows the generation of highly accurate Digital Elevation Models (DEMs) of the terrain or vegetated surfaces. The SAR data

(a)

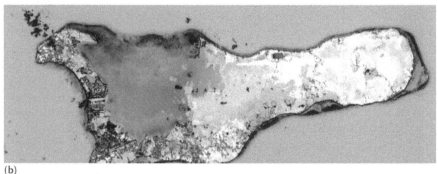
(b)

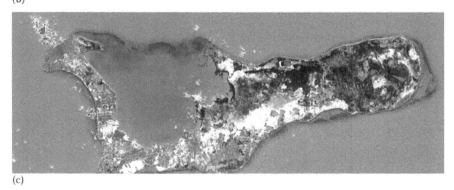
(c)

FIGURE 11.4
July 16, 2009, SPOT image of Grand Cayman classified using principal component analysis as the (a) first, (b) second, and (c) third principal components. (Data courtesy of ESA/CNES.)

itself can be used to interpret not only the canopy properties but also mangroves from other flooded vegetation that have well-pronounced microwave signatures (Proisy et al., 2000). For example, L band SAR data, alongside Shuttle Radar Topography Mission (SRTM)-DEM and ICESat-GLAS data, form the basis of the Global Mangrove Watch (Lucas et al., 2014) that's part of JAXA's Kyoto and Carbon Initiative. A revised global mangrove extent for 2010 was 140,260 km² (Bunting et al., 2022).

A, perhaps, surprising fact is that mangroves sequester more carbon per hectare than tropical forests, with the carbon being primarily stored below rather than above ground. In addition, it's not only mangroves that store carbon but also salt marshes and seagrasses that cover approximately 500,000 km² globally (International Blue Carbon Initiative, 2022).

11.3.4 Coastal Evolution, Including Sediment Transport

Climate change is expected to increase the frequency and magnitude of flood and storm events. Landslides are being triggered by intense or persistent rainfall, waves caused by storminess are making soft coastlines more prone to erosion, and higher sea levels together with more frequent storms are changing the way sediment moves through the oceans and around coasts. With more than a quarter of the world's population living close to the coast, society's vulnerability to these impacts is increasing.

According to Gillis (2014), 75%–90% of the Earth's natural sand beaches are shrinking due to increased storm activity, rising sea levels, and human development of the shoreline. As an example, the United Nations Environment Programme (UNEP, 2011) estimated that over the last 40 years, Jamaica's Negril beaches had experienced average beach erosion of between 0.5 and 1 m per year.

In many areas, beach replenishment is a major activity, either because the beach forms part of the protective barriers for the land or because the beach is a tourist attraction. However, sand is not an infinite resource, and most replenishment comes from other beaches, dredging, or mining: sand is one of the most consumed natural resources on the Earth, with the biggest user being the construction industry in the production of concrete. Therefore, EO monitoring of its extraction can support its conservation; see Lavender (2021) for a pilot project in Kenya and UNEP (2022) for a report on sand and sustainability consolidating from different sectors.

As discussed, this transport of sediment, such as sand, away from the beaches can be removal by human activities. Still, there will also be the seasonal and tidal processes where strong currents pick up the sediment and move it in plumes along the coast or out to sea. Quantifying these processes requires numerical models, run using climatic and meteorological data, combined with remote sensing data that supply the surface distribution of the sediment. Together, they can be used to quantify sediment fluxes throughout the water column.

Figure 11.5 is a Landsat-8 image showing the mouth of Chesapeake Bay in Maryland, USA, displayed as a pseudo-true-color composite (combining the red, green, and blue wavebands) with some enhancements to bring out specific features. The Chesapeake Bay Bridge–Tunnel can be clearly seen stretching to the north, with several boats passing through it.

FIGURE 11.5
Landsat-8 image of Chesapeake Bay from February 28, 2014. (Data courtesy of U.S. Geological Survey.)

The sediment transport around the coast can be seen by the complex color patterns of turbulence and movement, and during storms, this sediment will include the larger, and heavier, sand particles.

11.4 Practical Exercise – New York Bight

This chapter's practical will explore preprocessed data sets; MODIS Level 2 (L2), OLCI L2, and Level 3 (L3) ocean color data alongside Landsat-8 Level 1 (L1) data. The ocean color data sets are much larger data sets than those used in the previous practicals. We're going to break it down into three stages, and as the L2 and L3 data sets are large, we'll be using a couple of facilities within SNAP to keep the processing power to a minimum. As usual, we'll take you step by step through each practical stage.

11.4.1 Stage One: Importing and Processing MODIS L2 Data

11.4.1.1 Step One: Downloading MODIS L2 Data

We're going to be using MODIS data for this practical, which was also used in Chapter 8 urban environment practical, but this time, we're going to use MODIS L2 ocean color data sets. In addition, rather than downloading via GloVis or EarthExplorer, the data sets will be acquired from the

NASA Oceancolor website (https://oceancolor.gsfc.nasa.gov/). This website requires a NASA Earthdata Login, which you previously created to download data from EarthExplorer in Chapter 8.

Begin by going to the website homepage, and then using the top menu bar, select the menu item Data > Browse Data > Level 1&2 Browser that takes you a very busy download screen with lots of different sections, but don't worry it is not as complicated as it looks! There is also a help button on the top right if you need further questions answered.

In terms of the screen itself:

- Top left is the data set selection with several options.
- Top middle and right are the geographic location selection with a map and some location selectors.
- Bottom of the screen is the date selection with a month and year matrix, together with detailed monthly calendars for the selected period.

The data for this practical involve the New York Bight area on August 06, 2022, acquired by both the MODIS satellites; Aqua is generally preferred to Terra for ocean color applications because it has a better-defined and more stable calibration. However, from 2014 onward, Aqua also began suffering from age related degradation.

To get the MODIS data, make the following selections:

- Data set
  - In the top left section of the screen, you'll notice the default selection is to have the MODIS Aqua data selected as indicated by the tick in the checkbox. If you wanted to choose more than one satellite, or another satellite, you'd select it here. As we only want to download Aqua data, there is nothing else to do in this top box.
  - All the other options can remain as their default settings.
  - Click the Reconfigure Page button just below the map in the middle. Note: If you don't do this, the web page won't remember this mission selection as the date and location are chosen.
- Date
  - From the month and year matrix, click on 2022 to select the year. The months of 2022 will be highlighted in yellow, and all the other years and months will not be highlighted. Next select August, which will remove the highlight from the early months of 2022 and leave just August in yellow. To the right, the detailed calendar for August will be highlighted in green.

- Select 06 August by clicking on the number six in the detailed calendar; you should get just the 06 August highlighted in green, and "Saturday, 06 August 2022" will be the title across the top of the map; the map will also have changed to show the multiple MODIS swaths collected on this day. Note: Be careful to ensure you only have 06 August shown. If eight days are highlighted in green and the words "8-day period beginning Friday, 5 August 2022" is across the top of the map, it means you clicked on the squiggle below the number, rather than the number six itself. Try again, clicking on the number six.
- Location:
  - On the map, select the New York Bight area of the US northeastern coast, by simply clicking on the area shown in Figure 11.6a.

You will be taken to a new page with files to download, thumbnail images for Quasi True Color, Chlorophyll and Sea Surface Temperature, with maps of the chosen location. We will need to download both the Ocean Color and the Sea Surface Temperature files. The first file, Aqua Ocean Color, has the filename AQUA_MODIS.20220806T175500.L2.OC.nc

This file name is the entity ID for the MODIS files where

- The first set of characters indicates the satellite; MODIS-Aqua, although it is reversed in the name separated by an underscore.
- After the dot, the following eight characters are the YYYYMMDD of the date of data acquisition, in this case, 20220806.
- The T and the following six characters indicate the time in hours, minutes, and seconds; that Aqua acquired at 17:55:00 UTC.
- After the dot, the following two characters represent the data level; L2 which stands for Level 2.
- After the dot, the following few characters represent the product: OC for Ocean Color, IOP for Inherent Optical Properties, and SST for sea surface temperature. We'll use these short codes to distinguish the different data sets for the rest of this practical.
- After the dot, the last two characters indicate these are NetCDF files.

Click on the AQUA_MODIS.20220806T175500.L2.OC.nc hyperlink, which will take you to the NASA Earthdata Login screen. Login with the same credentials you created earlier in Chapter 8 for EarthExplorer. Once logged in, the OC file will begin downloading immediately. If you are not returned to the website automatically, click the "REDIRECT TO APPLICATION", or if that does not work, click the Back button on your browser until you reach the previous screen with the download hyperlinks on it. This time click on the hyperlink for the SST file AQUA_MODIS.20220806T175500.L2.SST.nc,

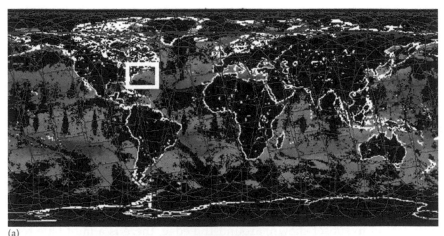

(a)

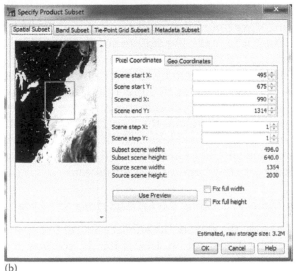

(b)

FIGURE 11.6
Downloading and importing New York Bight MODIS L2 into SNAP where (a) represents the position of the New York Bight on the NASA OceanColor Website and (b) is the New York Bight area of the MODIS image for subsetting. (Data courtesy of NASA.)

which is around 20 MB in size. Your login will be remembered, and the file will begin downloading. You should now have two files downloaded:

- AQUA_MODIS.20220806T175500.L2.OC.nc
- AQUA_MODIS.20220806T175500.L2.SST.nc

The .nc suffix indicates that these are NetCDF files created with internal compression so don't need to be unzipped.

11.4.1.2 Step Two: Importing the MODIS SST Data into SNAP

We're going back to SNAP for this practical. The first data set to import is the MODIS-Aqua SST file. Go to File > Open Product to bring up the file explorer window and then navigate to the SST file. Click on it, and press Return. The SST data set will be shown in the Product Explorer window. As noted at the start of the practical, these are large data files covering an extensive geographical area; therefore, we're going to subset the data before starting to work on it; this means you'll only be processing part of the scene that is, the New York Bight area, rather than the whole scene.

Click on the data set in Product Explorer window to select it, then go to the menu item Raster > "Subset", and the "Specify Product Subset" dialog box will appear. As we are subsetting geographically, we need to use the spatial subset tab, and there is a thumbnail version of the image we're about to subset shown; around the edge, there is a faint blue rectangle framing the image – it's the same shade of blue that's used in SNAP for zooming in. Move your cursor over one of the blue lines, or corners, left-click and drag it to make the square smaller; it's a bit fiddly, but try to create a rectangle to cover the approximate location shown in Figure 11.6b. Alternatively, we can help by giving you the Pixel coordinates we used, and you can use these for this practical. The coordinates are Scene start X: 495, Scene start Y: 675, Scene end X: 900, and Scene end Y: 1314. Once you've got your rectangle to the correct size and position. If you've selected the area using the box, take note of the coordinates you've used for later in the practical. Click OK. There will now be a second layer in the Product Explorer window beginning with subset.

11.4.1.3 Step Three: Processing the MODIS-Aqua SST Data

Click on the + sign at the start of the subset data set to expand it, and then do the same for the Bands option. Under Bands, you've got several products, including sst, qual_sst, and l2_flags.

Start by double-clicking on sst, which will open the SST image in the main window in black and white. On the Color Manipulation tab on the bottom left, ensure the Basic Editor radio button is selected. Then, click on the gray button at the end of the box under Palette to bring up all the color palette options. Scroll down and select the Spectrum color palette. Click the radio button for the Sliders editor, and perform a contrast enhancement to focus the stretch on the prominent histogram peaks; refer back to Section 6.10 for a reminder of the detailed instructions.

The result will be a map of SST in degrees centigrade, similar to Figure 11.1d shown earlier, where the lower temperatures show up as black/blue through to the warmer temperatures in orange/red colors. Looking at the histogram stretch, you'll see details on the units and what the different colors represent.

One of the bands in this data set is l2_flags, a series of masks that can be overlaid on the data to give various details and information about the processing that occurred. The masks are used by going to the menu item View > Tool Windows > Mask Manager (an icon for this is also on the toolbar), which will open a dialog box over the right side of the screen showing the image view. It lists a series of masks available for the SST data set, including the mask name, calculation method, color on the image, level of transparency, and description. Masks are turned on and off by clicking the checkbox next to their name.

At the bottom of the list are flags specifically related to SST that show the Best SST retrieval, Good SST retrieval, Questionable SST retrieval, Bad SST retrieval, and where there was No SST Retrieval. If you turn these checkboxes on and off, you'll see the different quality levels of the data superimposed on the SST view. It can be helpful to change the size of the Mask Manager dialog box by dragging the edge to make it narrower, which will enable you to see the whole SST view and where the masks appear. Close the Mask Manager when you have finished with it.

The qual_sst band, which you can open by double-clicking it, is essentially an amalgamation of all of these retrieval flags; the image shows values of between 0 and 4, where the 0 indicates the best quality, shown in black, through grays to white which is for 4 and indicates a complete failure or masked data, which is usually land.

11.4.1.4 Step Four: Importing and Processing the MODIS OC Data in SNAP

Repeat step two again using the Aqua data set to import and subset the OC data. This time, the original file size is 48 MB. This data set is a little harder to see in the subsetting thumbnail, which is why noting down the coordinates in step one can be helpful here. To subset the same location, enter the coordinates you used earlier; although it doesn't need to be precisely the same for this practical.

When you open the Bands for the subset of this data set, you'll see a wide array of available products that can confuse first-time users. The bands listed are as follows:

- R_{rs}: remote sensing reflectance at each wavelength, which we'll use to create a pseudo-true-color image.
- Diffuse attenuation coefficient (K_d): provides information on the rate of attenuation, or loss, of light within the ocean with depth.
- Aerosol optical thickness and Angstrom exponent are by-products of the AC and provide information on the aerosols in the atmosphere.

- Chlor_a for the chlorophyll-a concentration, which we'll use in this practical.
- Particulate inorganic carbon and particulate organic carbon are the particles of inorganic and organic material.
- Latitude and longitude arrays.
- Normalized fluorescence line height (FLH), which was described in Section 11.1.
- Instantaneous Photosynthetically Available Radiation (iPAR) and daily estimated PAR: the amount of downwelling solar irradiance as a single value covering the EM range from 400 to 700 nm. iPAR is for the time the satellite went over and is not calculated under clouds, while PAR is an estimate for the whole day and hence all the pixels (unless masked for another reason) have values.

We will start the ocean color processing by creating a pseudo-true-color composite. Go to the menu item Window > "Open RBG Image Window", and the RGB dialog box should be prepopulated with the profile NASA MODIS L2; wavebands R_{rs} 677, 531, and 443 are displayed as Red, Green, and Blue, where R_{rs} stands for the remote sensing reflectance. Select OK to create the composite. Then perform a contrast enhancement using the Sliders editor on the Color Manipulation tab for each waveband to produce a view similar to the earlier Figure 11.1b, showing the variations in color in these coastal waters.

Note:

- Although the blue band is a two-peak histogram, this is an ocean color image, and thus there are no pixels over the land; the two peaks represent the open ocean and coastal water pixels, respectively. Figure 11.1b shows a stretched version of the color of the water in the ocean, where the open offshore ocean is blue, and nearer to the coast, the yellow and red colors indicate that the water has a variety of SPM in the water column.
- There is a large area missing in the center of the image, which is sun glint – introduced in Section 11.1.2. It can significantly affect the accuracy of the values of the retrieved data and so is often masked. If you turn on the Mask Manager again, and select the MODGLINT flag, it will show the location of the sun glint matching the area of missing data.

Second, we will display the Chlor-a image, which is opened by simply double-clicking on the chlor_a option under Bands. Add the Spectrum color palette on the Basic Editor of the Color Manipulation tab. Switch to the Sliders editor to perform a contrast enhancement, which you can do by moving the triangles, as previously. Still, as the Chlor-a values range from very low to high, with both ends relating to structures in the water, it's common to apply a logarithmic scale to the histogram, which is done by pressing the Log_{10} button. This non-linear scaling should produce an image similar to Figure 11.1c, which shows the concentration of Chlor-a in the water, indicating that it increases closer to the shore with higher concentrations.

Like SST, there are masks that can be overlaid on the ocean color data, and they work exactly as for SST. For example, select the Chlor-a image, go to the menu item View > Tool Windows > Mask Manager, and go down the list to the CHLFAIL flag. Using the checkbox to turn it on and off, you may be able to see pixels around the coast where the Chlor-a algorithm failed. **Note:** It may be easier to see if the color of the displayed flag is changed; click on the color box for CHLFAIL, and you'll have a dropdown list of colors to choose from. Selecting white will make the pixels where the algorithm failed clearer.

11.4.1.5 Step Five: Download and Import the OLCI L2 Product

You can download the OLCI L2 data using the L1&2 Browser on the NASA Ocean Color website we used for the MODIS data. Go back to the NASA Oceancolor website home page, turn off the MODIS data by clicking on the checkbox to remove the tick, then select all four of the OLCI options for Sentinel-3A and -3B. Then press the Reconfigure Page button to see the OLCI rather than MODIS data; it will offer data starting in 2016. Select 2022, August, and then August 6, and the map title should now be "Saturday, August 6 2022". Finally, go to the map view and click somewhere in the New York Bight region.

You'll get taken to the download screen with hyperlinks for two data sets:

- OLCI L2 Ocean Color file acquired by Sentinel-3A at 15:20 on August 06, 2022 called S3A_OLCI_ERRNT.20220806T152042. L2.OC.nc, and
- OLCI L2 Ocean Color file acquired by Sentinel-3B at 14:41 on August 06, 2022 called S3B_OLCI_ERRNT.20220806T144133. L2.OC.nc

We are only going to use the Sentinel-3B data set for this practical, as this has better coverage of the New York Bight area – although it is upside down, but you will see that shortly.

Alternatively, we can also use the Direct Data Access route. Go back to the NASA Ocean Color Home page and then follow the menu Data > Direct Data Access. From the directory structure, go down the levels to Level-2/

S3B-OLCI/2022/06 Aug 2022 (218) and choose the OLCI L2 Ocean Color file for 14:41 on August 06, 2022, called S3B_OLCI_ERRNT.20220806T144133. L2.OC.nc.

Open the product in SNAP using the File > Open Product menu, and navigate to the file you have just downloaded. Once in SNAP, open the product with the +, and then open the Bands, where you'll see a similar, although slightly different, set of files to MODIS. Load the Chlorophyll-a (chlor_a) band by double-clicking on it. Move to the Color Manipulation tab and apply the Spectrum color palette in the Basic Editor section. Move to the Slider Editor, and use a logarithmic scale to the histogram, which is done by pressing the Log_{10} button.

Unlike the MODIS data, the OLCI products are for a whole orbit; as seen from the long-thin shape of the loaded data. To make matters slightly more confusing, Sentinel-3B is on an ascending orbit trajectory, acquiring data from South to North, so the New York Bight is toward the bottom of the strip. If you go to the Mask Manager, and switch on the Land mask, it might be easier to find. If you zoom into the strip and scroll down the image, you should find an area that looks like Figure 11.7a. On the right of this image, just above the center in red, is the corner of Lake Ontario. Nova Scotia is near the bottom of the image view, and across to the left of Lake Victoria is New York; as shown in Figure 11.7b.

Finally, you need to subset this data set to cover a similar area to MODIS. Click on the name of the Sentinel-3B data set in the Product Explorer window to select it. Then go to the menu item Raster > "Subset", and the "Specify Product Subset" dialog box will appear. It is possible to use the blue rectangle to select the area, but remember that the region we want is three-quarters of the way down the image, and the thumbnail is in black and white. The best tip we can give you to do this is to look for Lake Ontario on the right of the thumbnail to orientate yourself.

Of course, we also can give you the coordinates; enter the following into the Pixel coordinates option: Scene start X: 0, Scene start Y: 6832, Scene end X: 1216, and Scene end Y: 10288 to select the area. Once you've got your rectangle to the correct size and position, click OK. There will now be another layer in the Product Explorer window beginning with subset; see Figure 11.7c and d for the full scene and area of the subset.

11.4.1.6 Step Six: Save the Products

The last step is to save the subsets created so that, in the next stage, they can simply be imported rather than recreated.

To save a product, click on the data set you want to save to select it, and then go to the menu item File > Save Product. A dialog box will open, indicating the product is about to be converted into BEAM-DIMAP format, and click Yes. A standard file save dialog box appears, where the filename

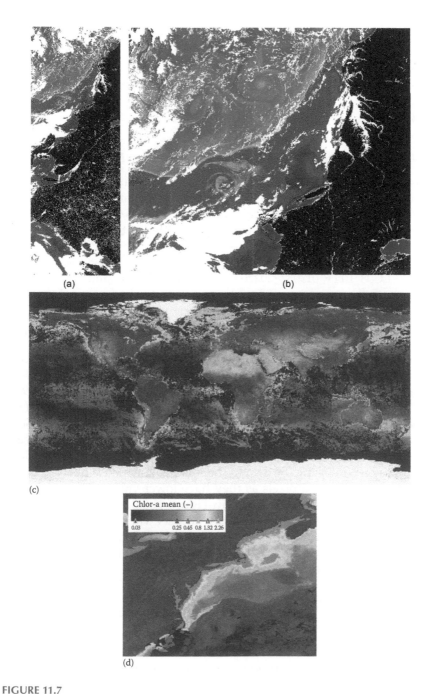

FIGURE 11.7
(a) Landsat-7 pseudo-true-color composite for September 8, 2002, (b) MODIS-Aqua pseudo-true-color composite reprojected into a UTM projection for the New York Bight, shown alongside the (c) global MODIS L3 8-day Chlor-a composite that's (d) zoomed in to show the New York Bight area. (Data courtesy of NASA/USGS.)

and its final location can be chosen. This approach creates a file with the extension ".dim" and a directory with the extension ".data" that contains all the bands as individual files. You should end up with three saved subsets, the MODIS subsets for OC and SST, and the Sentinel-3B subset.

11.4.2 Stage Two: Comparison of MODIS L2, OLCI L2, and Landsat Data

11.4.2.1 Step Seven: Restarting SNAP and Importing Landsat Data

This practical stage focuses on comparing Landsat and L2 MODIS/OLCI data, and seeing what's visible for the coastal waters at Landsat's 30-m resolution versus the 1-km resolution for the L2 MODIS/OLCI data. If you're continuing from the previous stage, the first step is to close down SNAP and reopen it to clear all the previous data sets.

Following the approach outlined in Chapter 8, download the Landsat-8 scene: LC08_L1TP_013032_20220806_20220817_02_T1 and then unzip. Then go to the menu item File > Import > Optical sensors > Landsat > "Landsat-8 in 30m (GeoTIFF)" and navigate to the LC08_L1TP_013032_2 0220806_20220817_02_T1_MTL.txt file; select it and then press the Import Product button.

To create a pseudo-true-color RGB image, open up the imported data set and select Bands, and now go to menu item Window > "Open RGB Image Window". Then select from the dropdown menu "Landsat8/9 L2 red/green/blue", although it may automatically be selected as the default option. Click OK, and the image will appear in the main window. Go to the Color Manipulation tab, and adjust the contrast enhancement for each waveband; where there are two histogram peaks, or a peak with a long tail; focus on enhancing the data in the first peak, which is likely to be the water rather than land pixels as they have lower values. The image will be noisy over the ocean, as shown in Figure 11.8a, as Landsat is a land-optimized instrument, although you'll still see various eddies, currents, and SPM in the coastal areas. As we saw in MODIS-Aqua, there was a strong influence of sun glint, which is also evident in the Landsat-8 image with sea surface features showing up strongly.

11.4.2.2 Step Eight: Importing the Previous OC Product

To import the Aqua OC subset product, saved from the previous session in BEAM-DIMAP format, go to the menu item File > Open Product. Then select the Aqua OC subset product, press the Open button, or return, and the subset version of OC will be imported into SNAP. Make sure you select the SNAP BEAM-DIMAP format single file, rather than the folder where the data are as they will both have the same name.

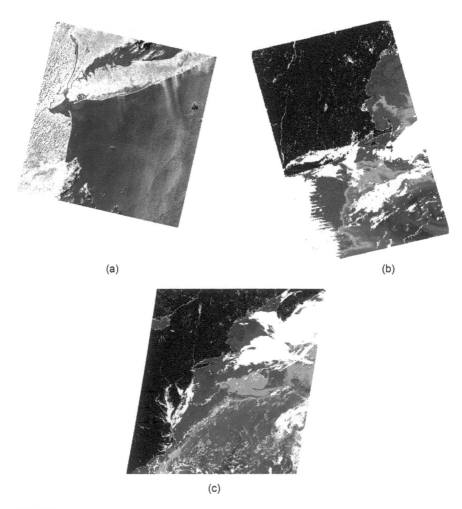

(a)

(b)

(c)

FIGURE 11.8
The New York Bight on August 06, 2022, as acquired by (a) Landsat-8 as a pseudo-true-color composite (b) MODIS-Aqua as a chlorophyll-a product, and (c) Sentinel-3B as a chlorophyll-a product. (Data courtesy of NASA/USGS/Copernicus/ESA.)

11.4.2.3 Step Nine: Reprojection of the OC Image

This stage of the practical is about comparing the Landsat and MODIS/OLCI images of the New York Bight. However, this isn't easy as the two data sets are in different projections. The Landsat data have already been reprojected to UTM, while the MODIS data are still in their satellite projection.

The MODIS data needs to be reprojected into a UTM projection, the same as Landsat, to create a better visual comparison. Open the MODIS subset product you have imported in the previous step, and click on Bands.

Go to the menu item Raster > Geometric > Reprojection, which opens the Reprojection dialog box. Ensure that the source product is the OC data set, and that you are happy with the Directory where the reprojection will be saved as a BEAM-DIMAP file. Switch to the Reprojection Parameters tab.

- Select the radio button for Predefined CRS in the Coordinate Reference System Section.
- Click on the Select button at the end of the Predefined CRS box.
- In the Filter box that appears, type "32618"; this should give you a single result of "EPSG:32618-WGS84/UTM zone 18N," which you select by clicking on it and then press OK.
- Press Run to create the reprojected file, and a new dialog box will open, telling you this has happened. Click OK and then Close to shut the reprojection dialog box.

Open the reprojected data set, and then expand the Bands. Load the Chlorophyll-a (chlor_a) band by double-clicking on it. Move to the Color Manipulation tab and apply the Spectrum color palette in the Basic Editor section. Then move to the Slider Editor and apply a logarithmic scale to the histogram by pressing the Log_{10} button; you can undertake a slight contrast stretch to remove the front tail of the image. Finally, go to the Mask Manager and turn on the Land Mask, and you should have a view similar to Figure 11.8b.

Repeat the same process for the Sentinel-3B subset: import the data, reproject it, add the color palette, undertake a logarithmic contrast stretch, and complete a final manual contrast stretch to remove the leading tail. Finally, go to the mask Manager and turn on the Land Mask, and you should have an image similar to Figure 11.8c.

The projected and reprojected versions of the MODIS and OLCI data can now be compared with the Landsat image. However, it would help if you remembered that they are all at different spatial resolutions, as Landsat is at 30 m, while MODIS and Sentinel 3 are at 1 km resolution. Therefore, to directly compare them, it is helpful to focus on a specific landmark. We'll use Long Island, which juts out from New York, and has the water body Long Island Sound above it and the Atlantic Ocean below it. On the Landsat view, Long Island is much larger in the upper part of the image and is placed toward the top right corner. On the MODIS image, it is much smaller, just to the left of the center. Finally, on the Sentinel-3 image, it is just above the left of the center. Comparing these three images can see how the different satellite instruments can show the same area slightly differently. The two OC-focused satellite data sets show the various levels of chlorophyll in the water, with currents and eddies visible.

This information is less obvious on the Landsat view, but with the higher resolution, you can see sediment leaving the mouth of the Hudson River.

Note: You'll need to switch back and forth between the viewing windows to compare the two images, which isn't the same as QGIS where you're displaying the layers on top of one another in the same viewer. This lack of stacked layers is a limitation of SNAP; therefore, even if you undertake image processing within SNAP, it can often be helpful to display the final results in QGIS.

11.4.3 Stage Three: OLCI L3 Data

11.4.3.1 Step Ten: Downloading OLCI L3 Data

For the final part of this practical, we're going to download and view an L3 image so you can see the advantages and disadvantages of using this level of data processing. Return to the NASA Ocean Color Home page and then the Data > Browse Data > Level 3 Browser.

You'll be presented with a series of thumbnail images of the world and a number of dropdown menu boxes to select the data. By default, the images show daily composites; all data for that day composited together at 4-km spatial resolution. There is also monthly data that have the advantage that there are far fewer holes, pixels with no data, because clouds are more prominent for the data composited with a shorter period, and there are more orbits included. However, dynamic features such as fronts or algal blooms can become blurred/smoothed by the compositing process. If you look at the 8-day and monthly options, even from the web page, you may be able to see fewer holes in the data with the more extended compositing period.

The variables you can choose from are the Product Status, Sensor, Product, Period, and Resolution. Using dropdown menus, by pressing the v at the end of each box under the title, gives you options, plus you can select a date range. For this practical, choose:

- Product Status: Standard. This is a standard product that the science community has broadly accepted; by changing to Provisional, Testing, or Special products from the first dropdown menu, a greater variety of options are presented with research products still under consideration or in development and testing
- Sensor: S3BOLCI that is the Sentinel-3B OLCI sensor.
- Product: Chlorophyll concentration.
- Period: Daily composite.
- Resolution: 4-km resolution.

- Start Date: 2022-08-06 to match the rest of the practical.
- End Date: 2022-08-06 to match the rest of the practical.

Select the thumbnail for the date of interest, August 6, 2022, that will be shown as the full screen. Hyperlinks to download in PNG file format are available at the bottom of the image at 4 and 9 km resolution. However, to download the full data set, you'll have two choices: SMI or BIN. The Standard Mapped Image (SMI) data are a rectangular pixel-gridded version of the BIN data. For this practical, choose the SMI NetCDF file to download, as SNAP does not recognize the BIN format. You may be required to accept another End User License Agreement, specifically for Sentinel-3, before you can download the data. This agreement includes the normal requirement for attribution and the relevant clause for any images produced using this data. Once acceptance is given, the data will be downloaded; it is a 10 MB file with the name S3B_OLCI_ERRNT.20220806. L3m.DAY.CHL.chlor_a.4km.NRT.nc.

As this is a NetCDF file, to load into SNAP you just need to go to File > Open Product and select the downloaded file as SNAP will recognize the format. If you expand the Product and then expand Bands you'll notice there is only one Band for chlor_a. Double-click on the chlor_a band to open it, showing you the whole world's image rather than just the New York Bight. It is in black and white, so go to the Color Manipulation tab and, in the Basic Editor, apply the Spectrum color palette. Move to the Slider Editor, and apply a logarithmic scale to the histogram; you should now have a set of brightly colored stripes representing the different orbit data acquisitions of Sentinel-3B; as shown in Figure 11.9a.

You may be able to tell where the New York Bight area is from this image. Still, it is helpful to add the land masses to the view. If you select the menu item Layer > World Map Overlay, you should see the land overlaid onto your image, and be able to identify the New York Bight more quickly. If you zoom in, you will get a view similar to Figure 11.9b; you can see the increased chlorophyll around the coasts and some offshore eddies in the waters.

The coarser resolution means, in terms of the level of detail, you can see a lot less than in the previous step. However, these L3 images help undertake large-scale or global research or reduce the amount of data to download and process by identifying potential trends at a larger scale and then investigating them with finer-resolution data. This approach can be enhanced by looking at one of the other data sets with a more extended time period that, as noted above, will fill in the gaps in the data set we were looking at. For example, if you download one of the monthly composites and add the color palette and logarithmic contrast stretch, you will end up with a Figure like 11.9c, showing the chlorophyll levels across the Earth.

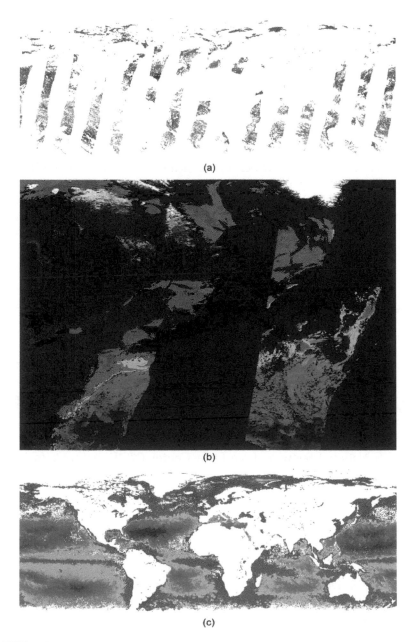

FIGURE 11.9
Global L3 Sentinel-3B 4 km resolution chlorophyll-a product for August 06, 2022, as the (a) whole world view and (b) zoomed in to the New York Bight overlaid with the SNAP World Map Overlay compared to the (c) global L3 July 2022 composite. (Data courtesy of NASA/Copernicus/ESA.)

11.5 Summary

This applications chapter has focused on the coastal environment, that is, both the terrestrial coastline that can suffer from erosion and the waters that contain microscopic phytoplankton. We've introduced the remote sensing field of ocean color in both theory and practical, and in addition, the practical has explored some of the different preprocessed data sets available.

If you're interested in the ocean, the skills learned here will allow you to research that environment and increase your understanding of how it's evolving, examine the water quality, and look at the oceanic parameters of temperature and salinity change. All of these elements are key components linked to climate change, and again you can begin investigating for yourself what is happening to the world.

11.6 Online Resources

- EUMETSAT Ocean and Sea Ice Satellite Application Facility (OSI SAF): https://osi-saf.eumetsat.int/products/sea-ice-products/
- Group for High-Resolution Sea Surface Temperature (GHRSST): https://www.ghrsst.org/
- NASA Oceancolor website: https://oceancolor.gsfc.nasa.gov/
- US National Snow and Ice Data Center (NSIDC): https://nsidc.org/data/amsre/index.html

11.7 Key Terms

- CDOM: Dissolved material in the water column resulting from land surface and subsurface, runoff, or biological degradation.
- Ocean color: The color of the ocean.
- Phytoplankton: Microscopic plants that have photosynthetic pigments such as Chlor-a.
- Remote sensing reflectance: Water-leaving radiance with variations in solar irradiance removed.

- Sea surface temperature: The water temperature measurement close to the ocean's surface.
- Suspended particulate matter: The organic and inorganic particles suspended in the water column.

References

Alongi, D. M. 2002. Present and future of the world's mangrove forests. *Environ Conserv* 29(3):331–349.

Bunting, P., A. Rosenqvist, L. Hilarides, R. M. Lucas and N. Thomas. 2022. Global mangrove watch: Updated 2010 mangrove forest extent (v2.5). *Remote Sens* 14:1034. https://doi.org/10.3390/rs14041034

Cavalieri, D., T. Markus, and J. Comiso. 2004. AMSR-E/Aqua daily L3 12.5 km brightness temperature, sea ice concentration, and snow depth polar grids, version 2 (updated daily). *National Snow and Ice Data Center, Boulder, Colorado.* Available at http://nsidc.org.data/ae_si12.html.

Clarke, G. L., G. C. Ewing and C. J. Lorenzen. 1970. Spectra of backscattered light from the sea obtained from aircraft as a measure of chlorophyll concentration. *Science* 167:1119–1121.

Duke, N. C., J. O. Meynecke, S. Dittmann et al. 2007. A world without mangroves? *Science* 317:41–42.

Fritz, H. M. and C. Blount. 2007. Thematic paper: Role of forests and trees in protecting coastal areas against cyclones, in chapter 2 protection from cyclones, report on Coastal protection in the aftermath of the Indian Ocean tsunami: What role for forests and trees? Available at http://www.fao.org/docrep/010/ag127e/AG127E07.htm (accessed April 17, 2015).

Gedan, K. B., M. L. Kirwan, E. Wolanski, E. B. Barbier and B. R. Silliman. 2011. The present and future role of coastal wetland vegetation in protecting shorelines: Answering recent challenges to the paradigm. *Clim Change* 106:7–29.

Gillis, J. R. 2014. Why sand is disappearing. *New York Times.* Available at http://www.nytimes.com/2014/11/05/opinion/why-sand-is-disappearing.html?_r=0 (accessed April 17, 2015).

Gower, J. F. R. and G. A. Borstadt. 1990. Mapping of phytoplankton by solar-stimulated fluorescence using an imaging spectrometer. *Int J Remote Sens* 11(2):313–320.

Green, E. P., C. D. Clark, P. J. Mumby, A. J. Edwards and A. C. Ellis. 1998. Remote sensing techniques for mangrove mapping. *Int J Remote Sens* 19(5):935–956.

Hallikainen, M., J. Kainulainen, J. Seppanen et al. 2010. Studies of radio frequency interference at L-band using an airborne 2-D interferometric radiometer. In *2010 IEEE International Geoscience and Remote Sensing Symposium (IGARSS),*

2490–2491. Available at http://ieeexplore.ieee.org/xpl/login.jsp?tp=&arnu mber=5651866&url=http%3A%2F%2Fieeexplore.ieee.org%2Fxpls%2Fabs_ all.jsp%3Farnumber%3D5651866.

Hu, C., Lee Z., and Franz, B.A. 2012. Chlorophyll-a algorithms for oligotrophic oceans: A novel approach based on three-band reflectance difference, *J Geophys Res* 117:C01011. https://doi.org/10.1029/2011JC007395

Hudson, H. 1868. A second voyage of Henry Hudson for finding a passage to the East Indies, by the north-east. In *Collections of the New-York Historical Society, for the Year 1809*, Vol. I, 81–102. New York: Riley.

International Blue Carbon Initiative. 2022. Available at https://www.thebluecar boninitiative.org/ (accessed August 26, 2022).

Lavender, S. 2021. EO4SAS white paper on using earth observation satellite data to support sustainable sand monitoring in Kenya. Preprint at https://doi. org/10.6084/m9.figshare.14604456.v1

Lucas, R., L.-M. Rebelo, L. Fatoyinbo et al. 2014. Contribution of L-band SAR to systematic global mangrove monitoring. *Mar Freshwater Res* 65(7):589–603.

Lyzenga, D. 1978. Passive remote sensing techniques for mapping water depth and bottom features. *Appl Opt* 17(3):379–383.

O'Reilly, J. E., S. Maritorena, B. G. Mitchell et al. 1998. Ocean color chlorophyll algorithms for SeaWiFS. *J Geophys Res* 103(C11):24937–24953.

Proisy, C., E. Mougin, F. Fromard and M. A. Karam. 2000. Interpretation of polari-metric signatures of mangrove forest. *Remote Sens Environ* 71:56–66.

Rast, M., J. L. Bezy and S. Bruzzi. 1999. The ESA medium resolution imaging spec-trometer MERIS—A review of the instrument and its mission. *Int J Remote Sens* 20(9):1679–1680.

Simpson, J. J. 1994. Remote sensing in fisheries: A tool for better management in the utilization of a renewable resource. *Can J Fish Aquat Sci* 51(3):743–771. http://dx.doi.org/10.1139/f94-074

Tian-Kunze, X., L. Kaleschke, N. Maaß et al. 2014. SMOS derived sea ice thickness: Algorithm baseline, product specifications and initial verification. *Cryosphere* 8:997–1018.

UNEP. 2011. The risk and vulnerability assessment methodology development project in Jamaica. Available at http://www.unep.org/disastersandcon flicts/Introduction/DisasterRiskReduction/Capacitydevelopmentand tech-nical assistance/RiVAMPinJamaica/tabid/105927/Default.aspx (accessed April 17, 2015).

UNEP. 2022. Sand and sustainability: 10 strategic recommendations to avert a crisis. GRID-Geneva, United Nations Environment Programme, Geneva, Switzerland. Available at https://www.unep.org/resources/report/sand-and-sustainability-10-strategic-recommendations-avert-crisis (accessed August 26, 2022).

12

Atmospheric Gases and Pollutants

The mixture of gases in the Earth's atmosphere is vital for life. Without the relatively thin atmospheric layer, compared with the diameter of the Earth, life would not exist on Earth as it protects us from harmful radiation from space and provides oxygen (O_2) we breathe. This O_2 is generated by plants, including those on land and in the sea. In this chapter, we are looking at how we can use Earth Observation (EO) data to understand what is happening in the atmosphere and how it impacts the air we breathe, in particular linked to air pollution and climate change.

12.1 An Understanding of the Atmosphere

The atmosphere is composed of a series of layers. The lowest layer, called the troposphere, extends from the ground to about 10 km and contains the air we breathe alongside the weather that influences our everyday lives. We are going to describe the troposphere in terms of three components: atmospheric gases, particulates, and water.

The atmospheric gases are in a relatively stable mix with around 80% nitrogen and 20% oxygen. Carbon dioxide (CO_2) and water vapor (H_2O) are found in much smaller proportions alongside pollutants such as nitrogen dioxide (NO_2). The CO_2 concentration in the atmosphere varies according to what is released and absorbed by processes occurring on Earth. As is now widely acknowledged, humans have been increasing the amount of CO_2 in the atmosphere through the release of carbon previously stored in other forms; for example, trapped underground in oil, gas, and coal deposits and within the vegetation such as trees. An increase in CO_2 causes the atmosphere to store more heat, which in turn causes global warming and leads to climate change. In addition, this CO_2 is absorbed by the ocean, so we also have an ocean that is becoming warmer and more acidic.

Solar radiation penetrates the troposphere easily, and the troposphere is also the layer that absorbs the heat reflected from the ground in a process called the greenhouse effect. Above the troposphere, there are several more layers, with the next layer being the upper atmosphere called the stratosphere. This layer contains ozone (O_3), often called the ozone layer,

DOI: 10.1201/9781003272274-12

a form of oxygen that prevents harmful ultraviolet solar radiation from reaching the Earth. Ozone is, perhaps, best known because of its impact on the stratosphere – the ozone hole in the ozone layer was in the news when it was depleted by chlorofluorocarbon and related halocarbon emissions that have since been curtailed with a ban in the 1980s. However, ozone is also generated in the troposphere when chemical reactions are caused by sunlight, for example, the breakdown of nitrogen oxides. Elevated concentrations can affect our ability to breathe by causing inflammation, particularly for people with respiratory diseases such as asthma. Over the last decade, urban background concentrations have been on an upward trend.

The second component of the troposphere is suspended particles. The concentrations vary according to what created them. A sandstorm is visible, but the particles are not suspended for long once the wind drops as they are heavy. When there are high concentrations of microscopic particles, which cannot usually be seen, they can become visible. For example, a haze over cities on calm days is caused by particles released from diesel and petrol cars alongside industrial and home sources.

The third component of the troposphere is water, which in its gas phase is water vapor. When the concentration is high and conditions right, this gas forms clouds and returns to the Earth's surface as rain and snow. When it rains, particles are removed from the atmosphere, and as a result, China purposely causes rain over Beijing (Guardian, 2020) which can suffer from high pollution levels. This change in city pollution also became evident during the COVID-19 pandemic when fewer people were using their cars, and once highly polluted cities started to have improved air quality. India's Central Pollution Control Board reported a 71% fall in nitrogen dioxide levels (Dasgupta, 2020). To have a look at further examples, the EO Dashboard (https://eodashboard.org/) developed by National Aeronautics and Space Administration (NASA), European Space Agency (ESA), and the Japanese Space Agency (JAXA) was originally created to support COVID-19 activities but has a much broader range of examples.

12.2 Detecting What Is in the Atmosphere

Atmospheric gases are highly individual in their ability to absorb and emit radiation, with each radiatively active gas having a specific absorption spectrum, that is, its own signature. Absorption of visible and near-IR radiation is primarily due to water vapor, ozone, and carbon dioxide. Given information about the vertical structure of the atmosphere and its composition alongside the viewing geometry, the spectrum of reflected solar radiation or emitted thermal radiation can be simulated using an atmospheric radiative transfer model. It is optical sensors and such models that EO missions use to calculate the gas concentrations (Matsunaga and Maksyutov, 2018).

A pioneering mission was the Scanning Imaging Absorption Spectrometer for Atmospheric CHartographY (SCIAMACHY) on board ENVISAT, which operated from 2002 to 2012. Although SCIAMACHY was not designed explicitly for greenhouse gas (GHG) measurements, it clearly showed the possibilities. It led to the development of the Copernicus TROPOspheric Monitoring Instrument (TROPOMI) launched on Sentinel-5P in October 2017. The advantage of EO data is that it provides unprecedented amounts of atmospheric composition measurements daily, with near-global coverage. Level 1 and 2 data are available from the Copernicus Open Access Hub (https://scihub.copernicus.eu/).

SCIAMACHY observed the solar radiation in the UV (10–400 nm) to short wavelength infrared (SWIR, 1,400–3,000 nm) regions of the electromagnetic spectrum under different viewing angles such as nadir (vertically down to the Earth's surface) and limb viewing (a slanted view through the atmosphere out to the Earth's horizon). Its spectral bands were carefully chosen to measure the ozone, nitric oxide, nitrogen dioxide, carbon monoxide, carbon dioxide, methane, and nitrous oxide concentrations.

Compared with SCIAMACHY, TROPOMI's design supports smaller ground pixels and shorter revisit times. TROPOMI aims to map emissions globally every 24 hours, taking measurements every second of an area of approximately 2,600 km wide and 7 km long at a resolution of 7 km.

Further optical missions include the MetOp series of three polar-orbiting meteorological satellites, the GOME-2 (Global Ozone Monitoring Experiment–2) instrument, plus MODIS and Copernicus Sentinel-3 which are used for quantifying the aerosols and water vapor. Water vapor is also quantified using global navigation satellites, such as GPS, as its effect significantly delays their radio waves.

12.3 Air Quality

Many cities suffer from poor air quality due to both local sources (e.g., urban traffic) and regional sources (e.g., heavy industry, dust storms, and forest fires) that can significantly affect human health and lifestyle. The World Health Organization (WHO) provides recommendations on air quality levels, as well as interim targets for six key air pollutants: particulate matter (PM), ozone, nitrogen dioxide, sulfur dioxide, and carbon monoxide. The small (less than 2.5 μm, PM2.5) particulate fraction varies substantially between and within regions of the world, with more than 90%

of the world's population living in areas where concentrations exceeded the agreed maximum level of 10 μg/m^3 in 2019, with the lowest levels being in the Americas and Europe (WHO, 2021). The WHO estimates that around 80% of deaths attributed to PM2.5 exposure could be avoided if countries reduced concentrations to below the recommended annual levels.

Satellite data can be used to measure aerosols, which is the variable satellites use to determine the concentration of particles in the atmosphere. The equivalent of the satellite aerosol product from models are the PM products that often cover different particle size fractions; for example, PM1 would be less than 1 micron (μm), PM2.5 for less than 2.5 μm, and PM10 for less than 10 μm. These microscopic particles can be derived from human activities, such as fires and diesel cars, and natural activities, such as pollen, and can travel large distances in the air. In addition, particles less than 2.5 μm in size may enter the human body, including the bloodstream, and can affect health. Therefore, they are monitored as part of understanding air quality.

Most of the current aerosol retrieval algorithms are devised for global monitoring, such as the standard aerosol product of NASA (MOD04), which means that, within urban areas, the relatively low spatial resolution results in an inability to separate the aerosols from the brighter and more variable targets, such as buildings, which cannot be easily characterized. Errors of up to 20% were reported by Hsu et al. (2004) for products derived from Sea-viewing Wide Field-of-view Sensor (SeaWiFS) and MODIS using a dark target-focused algorithm (Levy et al., 2007) because this approach didn't cope well with bright targets. Since then, the MODIS algorithm has been updated to combine the approach proposed by Hsu et al. (2004) with the dark target algorithm.

Bilal et al. (2014) used MODIS products at 500 m spatial resolution to derive aerosol concentrations over Beijing linked to dust storms, which included an extreme floating dust storm on April 17, 2006, originating in the Gobi Desert (Logan et al., 2010; Lue et al., 2010), that led to severe air pollution. Figure 12.1a shows the MODIS top-of-atmosphere reflectance for this dust storm as a pseudo-true-color composite using wavebands 1 (centered at 645 nm), 4 (555 nm), and 3 (470 nm) as red, green, and blue; the data displayed have a resolution of 1 km. It shows the dust as a brown mass over both the land and Bohai Sea region, which is coloring the cloud brown where it overlaps. Alongside Figure 12.1b is the MODIS MOD08_D3 product (Platnick et al., 2015) containing L3 gridded daily aerosol, water vapor, and cloud values configured to show the combined land and ocean aerosol optical depth with high values associated with the dust aerosols, shown as the orange and red colors, in the Beijing/Bohai Sea region for this same date. This image isn't showing aerosol concentrations; instead, it's the linked variable optical depth that shows how quickly light is absorbed in the atmosphere, with higher values equating to higher aerosol concentrations. However, the coarser spatial resolution of the MOD08_D3 product, of 1° of latitude and longitude, prevents the details for the Beijing/ Bohai Sea region from being visible, although high values are shown.

The L1 and Atmosphere Archive and Distribution System (LAADS) website (http://ladsweb.modaps.eosdis.nasa.gov) provides the route for downloading MOD08_D3 and other products. In the practical exercise in Chapter 11, aerosol products were reviewed as a by-product of performing an atmospheric correction.

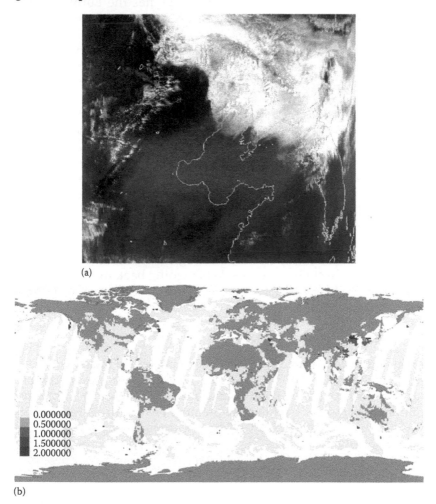

(a)

(b)

FIGURE 12.1
MODIS data from April 17, 2006, shown as the (a) MODIS Top-of-Atmosphere reflectance as a pseudo-true color composite using wavebands 1, 4, and 3 over Beijing and the Bohai Sea alongside the (b) global combined land and ocean aerosol optical depth from the MOD08_ D3 product. (Data courtesy of NASA.)

12.3.1 Real-Time and Forecasted Alerts

The large-scale emission of pollutants by activities such as forest fires can cause a significant air quality hazard. On such occasions, EO data are used to locate the sources and track and quantify the emissions, requiring that the EO data are available in real time. Satellites used for forest fire monitoring include MODIS, VIIRS, and Sentinel-3. NASA has the Fire Information for Resource Management System (https://firms. modaps.eosdis.nasa.gov/map/) while Europe has the European Forest Fire Information System (https://effis.jrc.ec.europa.eu/); both generate global maps.

In Europe, the Copernicus Atmosphere Monitoring Service (CAMS) combines Copernicus EO data with an ensemble of state-of-the-art numerical air quality models to provide both a real-time and forecast service regionally and globally. The real-time forecasts are available on the CAMS website (https://atmosphere.copernicus.eu). CAMS pulls together various measures of atmospheric gases and pollutants and uses these to create an overall assessment of air quality by the hour and day. There are also maps of specific parameters, including ozone, nitrogen dioxide, sulfur dioxide, carbon monoxide, and aerosols. While it also provides daily global air quality modeling for specific parameters, including carbon monoxide, aerosols, and UV radiation.

In terms of the global reanalysis, historical data from CAMS will be used in the practical exercise. The advantage of reanalysis data is that it does not have the constraint of issuing timely forecasts, so there is more time to collect observations. When going back in time, improved versions of the original observations can be ingested, improving the final product's quality.

12.3.2 The Impact of COVID-19

Nitrogen dioxide is a toxic gas from the combustion of fossil fuels in vehicles, power plants, and factories. It has been linked to respiratory problems and other health conditions. It also reacts with other chemicals in the atmosphere to form fine particulate pollution, which the WHO identified as the leading cause of the Earth's 7 million annual deaths from air pollution (WHO, 2014).

Sentinel-5P's TROPOMI saw a greater than 50% drop in this air pollution within European cities due to the impact of the coronavirus lockdowns; Paris saw a 54% drop in nitrogen dioxide, while Madrid, Milan, and Rome saw a decline of nearly 50% (McMahon, 2020). As a further example, Figure 12.2 shows the average nitrogen dioxide concentrations from March 25 to April 20, 2019, and average concentrations from the same date range in 2020. The spikes in the top image show concentrations

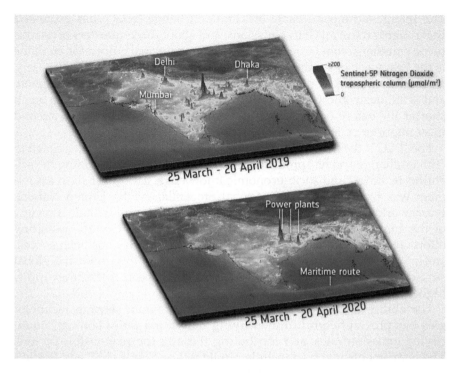

FIGURE 12.2
Copernicus Sentinel-5P data (2019–2020) showing Nitrogen Dioxide levels (processed over India by ESA, CC BY-SA 3.0 IGO).

from 2019 over Delhi and Mumbai. However, in contrast to the 2020 image taken during COVID-19 restrictions, the peaks for cities have disappeared, while the high concentrations in northeast India coincide with the ongoing activity in coal-based power plants in east India. There is also a persistent signal from shipping (maritime route).

12.4 Greenhouse Gas Emissions

GHGs from human activities strengthen the greenhouse effect, warming the Earth due to trapped heat, which in turn causes climate change. The most well-known GHG is carbon dioxide, but there are other gases such as methane and nitrous oxide, also known as laughing gas. Nitrous oxide is about 300 times as potent as carbon dioxide at heating the atmosphere, and it is long-lived and contributes to depleting the ozone layer.

The Intergovernmental Panel on Climate Change (IPCC) has estimated that roughly 6% of all GHG emissions, and about three-quarters of nitrous oxide emissions, come from agriculture. The principal culprit for methane is the heavy use of synthetic nitrogen fertilizer (Chrobak, 2021). Similarly, methane is more potent than carbon dioxide at heating the atmosphere and is famously known as the gas from livestock flatulence. It has a much shorter life before being broken down but is still classed as the second most important GHG.

The IPCC is the United Nations body for assessing the science related to climate change and has produced a series of reports summarizing the scientific, technical and socio-economic knowledge. The latest sixth assessment was to inform the 2023 Global Stocktake by the United Nations Framework Convention on Climate Change, which will include a review of the goal to keep global warming to well below 2°C while pursuing efforts to limit it to 1.5°C. Each of the last four decades has been successively warmer than any decade preceding it since 1850, with the global surface temperature during 2001–2020 being between 0.84°C and 1.10°C higher than 1850–1900 (IPCC, 2022).

The ability to map GHG emissions using remote sensing technologies has proven helpful in identifying individual point sources, quantifying emission rates, and attributing these to the responsible sectors. The instruments involved include TROPOMI on Sentinel-5P and NASA's Orbiting Carbon Observatory (OCO) -2 onboard a satellite and OCO-3 attached to the International Space Station (ISS). The original OCO satellite was launched in 2009 but was lost due to a launch vehicle failure, while OCO-2 was launched in July 2014 and OCO-3 was deployed in May 2019. Both active OCO spectrometers include three spectral wavebands centered on the molecular oxygen-A band at 765 nm and the two carbon dioxide bands at 1,610 and 2,060 nm (Taylor et al., 2020). The derived products include solar-induced fluorescence emitted during plant photosynthesis, which is important as plants uptake carbon dioxide during photosynthesis, and column-averaged carbon dioxide in the atmosphere that can pinpoint specific emissions from human activities and sources such as volcanic plumes. Analysis of the data from OCO-2 concluded that the 2015–2016 El Nino-related heat and drought impacts in tropical regions were responsible for a record spike in global carbon dioxide (NASA, 2017). Further information on the OCO missions, including access to the data, can be found at https://www.nasa.gov/mission_pages/oco2 and https://www.nasa.gov/mission_pages/oco3.

Additional missions include the China National Space Administration's TanSat mission, Japan Aerospace Exploration Agency (JAXA) GHG observing satellites GOSAT and GOSAT-2, NASA Ozone Monitoring Instrument (OMI), and the China Aerospace Science and Technology Corporation's Terrestrial Ecosystem Carbon Inventory Satellite (TECIS).

In addition, there is a planned Copernicus Anthropogenic Carbon Dioxide Monitoring (CO2M) constellation.

12.4.1 Observing Methane

Methane is emitted from various sources, including livestock, as previously mentioned. In addition, oil–gas systems, landfills, coal mines, wastewater management, rice cultivation, and wetlands are considered significant sources of methane emissions. The atmospheric concentration of methane has risen since pre-industrial times, with human sources currently accounting for approximately 70% of total annual emissions.

Emission inventories used for climate policy rely on "bottom-up" estimates of activity rates and emission factors for individual source processes. "Top-down" information from observations of atmospheric methane is often at odds with these estimates, and differences need to be reconciled.

Landfill sites can provide ideal conditions for methane generation, making some sites potent sources and, if uncontrolled, emissions. As sewage sludge and agricultural waste may also be incorporated into landfills, the methane source strength for these categories may overlap somewhat. However, for much of the world, it is methane derived from the anaerobic decomposition of municipal rather than agricultural waste that dominates. Another large emitter is the energy sector, including oil, natural gas, and coal. This release can be through venting, where the gas is vented to the atmosphere, or flaring (burning) to prevent pressure build-up in coal mines/refinery plants/pipelines.

Atmospheric methane is detectable by its absorption of radiation in the SWIR and thermal infrared (TIR) around 8,000 nm. Unlike other instruments, such as TROPOMI, the commercial GHGSat constellation focuses not on global coverage but on specific targets with a much higher spatial resolution of around 50 m. In addition, planned geostationary instruments in the SWIR or TIR would allow a combination of high spatial and temporal resolution over continental-scale domains. The International Energy Agency provides data, analysis, and solutions on all fuels and technologies and has a "Methane Tracker" (https://www.iea.org/articles/global-methane-emissions-from-oil-and-gas) that uses EO data to provide an understanding of emissions across more than 70 countries. As they say, a key advantage of satellites is that they can help locate significant emitting sources promptly, and once a leak has been found, it can often be fixed relatively quickly. For example, Figure 12.3a shows the emission of methane linked to a gas pipeline using Sentinel-5P with the data processed by Kayrros; ESA (2021), and Figure 12.3b shows the same area as a pseudo-true-color composite with Sentinel-2.

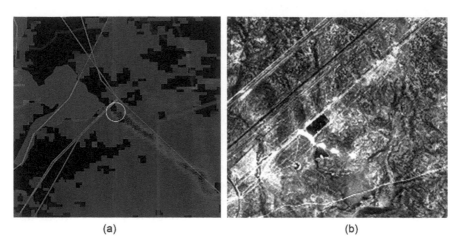

(a)   (b)

FIGURE 12.3
Methane hotspots over a gas pipeline in Kazakhstan detected by (a) the Copernicus Sentinel-5P mission and (b) Copernicus Sentinel-2 mission; modified Copernicus data (2020), processed by Kayrros, full article ESA (2021).

12.5 Practical – An Assessment of Air Quality and Temperature

For this practical, we will use QGIS to visualize the data available through CAMS and the Copernicus Land Service.

Unlike the previous exercises, these data aren't the original Level 1 or 2 satellite data, instead, it is data that has already been processed and is made available as a time-consistent data set. This form of data is essential to help distinguish real environmental changes from artifacts/issues caused by the data or its processing. In addition, unlike the previous exercises, we will not always be downloading all the data to our local computer. Instead, for one data set, we will connect directly to it over the Internet and get the data in real time. This means that you need to be connected to the Internet throughout this practical, to see the CORINE Layer when you want.

12.5.1 Stage One: Adding Cartographic Layers

As the first step, we will add some base Cartographic Layers into QGIS as we did in Section 7.8. Again, we are going to download the layers from the Natural Earth website (https://www.naturalearthdata.com/), and this time we want two layers:

- Populated Places layer from the cultural section of the small-scale 1:110 m group
- Coastline layer from the physical section of the medium-scale 1:50 m group

Unzip the downloads to get the files, then import the shapefiles into QGIS for each of the two layers using the menu item Layer Menu > Add Layer > Add Vector Layer. Full details on this process are in Section 7.8. Go to the Layer Properties on the Populated Places layer, and select the Layers tab. Change the No Labels dropdown item by pressing the downward-facing triangle at the end, and select Single Labels, then OK. This step will cover the map with city names.

Zoom into Europe and move the map so the capital of Germany, Berlin, is in the center of your screen. For using CORINE, in the next step, the scale of the layer on the screen will need to be 1:100,000 m or less. On the status bar at the bottom of the screen is the Scale box indicating what the current image is scaled at. To reduce it to 1:100,000 you can click the downward-facing triangle at the end of the Scale box and select 1:100,000. However, you can lose Berlin off your screen and will, hence, need to spend time moving around the white image to find it again. An alternative approach is to click on the zoom-in magnifier icon, and rather than clicking on the image, hold down the left mouse button and draw a box around your area of interest; in this case, Berlin. This approach will zoom in on the map view but keep Berlin visible. Repeat the exercise until your scale is approximately 1:100,000. You should end up with a screen similar to Figure 12.4a.

12.5.2 Stage Two: Adding CORINE Land Cover Data

For the second step, we will load the CORINE land cover data into QGIS as a Web Mapping Service (WMS). The CORINE data are available as a Pan-European layer, that is, they are only available for Europe. If you go to the CORINE page (https://land.copernicus.eu/pan-european/corine-land-cover), you'll see layers available for different years. We'll use the most recent at the time of writing, which is 2018, but you may find a more recent data set. If you click on the thumbnail of the layer you want, you should see a preview screen of the whole of Europe, with above it a button that says "Web services" at the top of the map. Click on this link, and a panel will appear to the left of the map preview listing the web services available, and there will be a link for CLC2018_WM. Click on this link, which will take you to a new page titled Corine/CLC2018_WM (MapServer). All you now have to do is look for the WMS hyperlink under the menus at the top of the screen and click on it; as shown in Figure 12.4b. This link takes you to another page with potentially confusing text as it

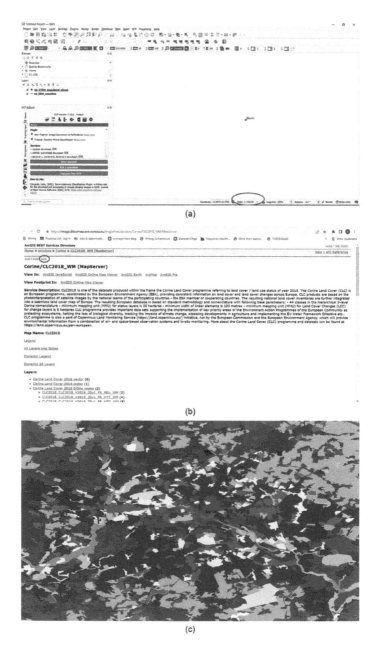

(a)

(b)

(c)

FIGURE 12.4
QGIS opens with (a) the Natural Earth 110-m Populated Places Berlin layer point from at the center of the screen and the location to set the scale at the bottom highlighted with an ellipse, (b) web page for the CORINE Web Mapping Service with the location to get the hyperlink in an ellipse, and (c) CORINE data around Berlin with the Berlin point location at the center. (Courtesy of the Copernicus Land Service/European Environment Agency.)

is a data set schema, but the great news is that you don't need to read this either! All you need to do is copy the full URL link at the top of the screen, which at the time of writing is https://image.discomap.eea.europa.eu/ arcgis/services/Corine/CLC2018_WM/MapServer/WMSServer?request =GetCapabilities&service=WMS

Now we'll go back to QGIS and add the layer. As this is a real-time web service, where data are being pulled from a server, rather than loaded from your local machine, you need to ensure that you are connected to the Internet before attempting to connect.

Go to the menu Layer>Add Layer>Add WMS/WMTS Layer, which brings up the Data Source Manager: WMS/WMTS dialog box. Click on New, and a Create a New WMS/WMTS Connection dialog box will appear. Give the new connection a name in the first box, such as CORINE, and in the second box, for the URL, paste the link you copied from the website above. Click OK, and you'll be returned to the Data Source Manager: WMS/WMTS dialog box, but the layer you have just created will be at the top of the Layers tab. Next click on Connect, and several layers will appear in the box underneath.

We'll select the "Corine Land Cover 2018 vector", so click on that layer, and when it is highlighted, click on Add, which should then Close. The Corine Land Cover 2018 vector layer will appear in the layer panel, and you should see the map covered in a myriad of bright colors representing the different types of land cover, similar to Figure 12.4c. If you click on the right-facing triangle to the left of the Corine Land Cover 2018 vector layer, a legend will appear underneath that lists all the different types of land cover available within the layer, and in what color they are represented.

Most of Berlin will be purple/pink/red colors equating to urban land cover types: dark red, continuous urban fabric (111): red, discontinuous urban fabric (112): purple, industrial or commercial units (121); pale pink, representing green urban areas (141); turquoise blue, water courses (511); pale blue, water bodies (512). To the city's left, you will see a large green area next to the water, the Forest Grunewald, and the river Havel extending in a southwest direction. On the eastern side of the city is the river Spree, which runs across the whole of Berlin, but the northeast part of the river is not very wide and so is merged into the broader land cover polygons that are classed as Industrial or commercial units.

You can move around the map using the hand icon and zoom in to see the details. However, if you zoom out too much, the layer will not appear because, as discussed earlier, the map scale for CORINE needs to be at least 1:100,000. When a sufficient zoom level is reached, the CORINE layer should reappear. This restriction limits the amount of data being pulled in real time across the Internet.

Although we are focusing on Berlin in this exercise, as this is a web service, the data for anywhere in Europe is available and accessible through this link. You can explore the land cover of the whole of the European continent by simply moving around the map. If this had been a downloaded data set, it would have been vast and cumbersome to download, and store. Web services are a developing feature of EO, and we'll talk more about them in Chapter 13.

12.5.3 Stage Three: Downloading the CAMS Data Set

The second data set we will use in the practical is from CAMS, a service developed by the European Union's Copernicus Program. From the Home page (https://atmosphere.copernicus.eu), click on Data in the top left, and then the button "Access the Atmosphere Data Store".

This hyperlink takes you to the site to obtain the data, but to get data, you need to have an account. It is easiest to do this before you start searching for data. The Atmosphere Data Store account, like all the services we are using in the book, is free to create and only requires minimal information.

Select the Login/Register button in the top left corner, select the Create New Account tab, or use the Register for free hyperlink. You'll need to enter an email address, password, and some details about yourself. After submitting your details, you are immediately logged into the Atmosphere Data Store.

To download the correct data, you need to know the location you are interested in, so go to QGIS. Then, click the mouse picture next to Coordinate at the bottom of the screen, which gives you the Latitude/ Longitude coordinates for the top and bottom corners of the view on your screen. For us, although yours may be different depending on where you put Berlin, it is

13.0753, 52.3074: 13.7865, 52.7018

The CAMS data have a coarse spatial resolution and so to ensure that you get the data you need, it is necessary to expand the area to select. We would suggest something similar to

`12.0000, 51.0000: 15.0000, 54.0000

Now we have our location, we need to navigate to the data set on the CAMS website. Once you are logged in, the home page of the Atmosphere Data Store has a search box in the center. Type "CAMS global" into the search box and press Search. You'll get several data sets offered, but scroll down until you find the CAMS global reanalysis (EAC4) data set, and click on the hyperlink.

This hyperlink takes you to a screen with details of this data set. EAC4, which stands for Centre for Medium-Range Weather Forecasts (ECMWF) Atmospheric Composition Reanalysis 4; it is the fourth generation of the ECMWF global reanalysis for atmospheric composition. This time-series data set is the output of a computer model that has ingested actual in situ observations and satellite data with new global estimates of the atmospheric composition every 3 hours. EAC4 is helpful because rather than taking data from a single satellite, or even multiple satellites, the model data set is continuously available from 2003 to the present day and provides several consistent parameters from a single source. This previous generation of a time-series data set is a huge advantage when looking at how things change over time. Select the Download Data tab from this screen, and you'll now see a series of options.

- Variable: Click on the downward-facing triangle next to Single level and select the variables:
  - 2 m temperature: air temperature 2 m off the ground
  - Particulate matter $d < 1$ μm (PM1): the particles that are less than 1 μm in diameter
  - Particulate matter $d < 10$ μm (PM10): the particles that are less than 10 μm in diameter
  - Total column nitrogen dioxide: total amount of NO_2 throughout the whole height of the atmosphere
  - Total column ozone: total amount of O_3 throughout the whole height of the atmosphere
- Date: Enter a range of 2018-01-01 to 2021-12-31 for a 4-year data set.
- Time: 12.00 to get the data at midday.
- Area: Select the radio button for Restricted Area, and then enter the four expanded coordinates from QGIS above; namely 12.0000, 51.0000: 15.0000, 54.0000, with the first pair being west and south, respectively, and the second pair being east and north, respectively.
- Format: Select the NetCDF format.
- Accept the License Terms: Again, this confirms that the data can be used free of charge for any lawful purpose, so long as it is correctly attributed and the appropriate attributions are shown within the terms.

Once you have completed the requirements, press the green "Submit Form" button. The file will be available instantaneously, so click the Download button to download the small 374 KB file. Once downloaded, we'd suggest you rename the file as something like "CAMSBerlin.nc" as the download name itself is a very long combination of letters and numbers that will mean nothing to you!

The downloaded data are different from those we've used before. It is pixel data that will be stored in an unstructured grid format because models don't necessarily have square pixels, which means the data set needs to be loaded into QGIS as a mesh layer. To load the data, go to menu Layer>Add Layer>Add Mesh Layer menu to open the Data Source manager mesh, use the ... to navigate to the file you have just downloaded, select it and press Open, then click Add and you will see the data appear as a layer at the top of the layers window. Click Close.

If Berlin has disappeared, ensure that the ne_110m_populated_places layer is at the top, with the ne_50m_coastline as the second layer, CAMS Berlin as the third layer, and Corine Land Cover 2018 as the bottom layer. Your main image is also likely to have gone a single color because the CAMS data are at a much coarser resolution than the Corine data, and you will need to zoom out to see it correctly. However, before you do that, we need to adjust how the data are displayed. If you right-click on the new layer, select Properties, navigate to the Symbology tab, and you'll see five further tabs. The first one, which has a hammer and screwdriver icon, is the Data sets tab and lists all the variables you downloaded. The 2 m temperature is being shown, as indicated by the color palette symbol in color, rather than the other four layers that have the color palette symbol in gray.

To adjust the color palette, click on the color palette tab that is the second one, and then:

- Click on the downward-facing triangle at the end of Color ramp, then on the right-facing triangle next to All Color ramps. From this list, select the RdBu color ramp. The higher numbers are blue, and the lower numbers are red.

- Click on the downward-facing triangle at the end of Color ramp again, and this time click on Invert Color ramp. All this does is change the direction of the values, so higher (warmer) values are in red, and the lower (colder) values are in blue, which is the more familiar. Note: The inversion does not always hold, particularly if you save the QGIS project and go back in, so you will need to check on the range in the Layer panel that the red is at the higher values.

- Adjust the Minimum and Maximum Values of the color palette so the contrast stretch is over the range 270.0–290.0 K.

- Scroll down the color palette, past the table showing the associated values and colors. Underneath that is an option for Mode. Click on the downward-facing triangle to give you the options, and switch from Continuous to Equal Interval. Then go to the right of the box, and change the number in the Classes box from 52 to 9, which is the number of intervals in the color palette.

- Click OK to apply all these changes.
- Turn off the Corine layer by clicking on the tick box.

If you now Zoom out, you will eventually get to an image similar to that shown in Figure 12.5a; a mostly light pink box. However, a slightly darker layer runs across the middle and up the right edge, indicating that this area is warmer than the rest of the image. We admit that it is not the easiest to see, unless you have a fantastic ability to distinguish between shades of light pink! Instead, if you go to the layer Properties, Symbology tab, on the color palette tab, press the downward-facing triangle on the Interpolation option, then change the option from Linear to Discrete, and click OK. You should now get a more apparent highlighted area showing the higher temperatures, as in Figure 12.5b. This highlighted area corresponds reasonably well to the urban center areas of Berlin, and if you switch the Corine Layer back on and zoom in, you will see the comparison. To see more detail than available with CAMS, you could download data from the Copernicus Climate Change Service (C3S), again part of the European Union's Copernicus Program, which holds the ERA5 data set that is the fifth generation of the ECMWF atmospheric reanalysis of the global climate. This data set has a much smaller spatial resolution; you can see an example image over Berlin in Figure 12.5c.

We didn't include it within the practical as the data set is huge, and a single month equates to around 3 GB of data when unzipped, meaning that downloading a time series would take a lot of storage space. However, if you have the capacity and want to download the data, it can be found at the Copernicus Climate Data Store (https://cds.climate.copernicus.eu/#!/home). As with the Atmosphere Data store, you'll need to create a free account to download data. Figure 12.5c was generated using data from the "Climate variables for cities in Europe from 2008 to 2017", which is the 2 m temperature data for 100 European cities. Still, you can also download several other variables, including PM, nitrous oxide, and ozone. Once downloaded, the ERA5 data will also be loaded as mesh data into QGIS in the same way as the CAMS data.

12.5.4 Stage Four: Visualizing the CAMS Time Series

The downloaded CAMS data set covered 4 years, so it can be helpful to plot the data as a time series, which helps to identify any patterns, changes, or anomalies over time. To see a time series plot of the CAMS data, a QGIS Plugin called "Crayfish" is required.

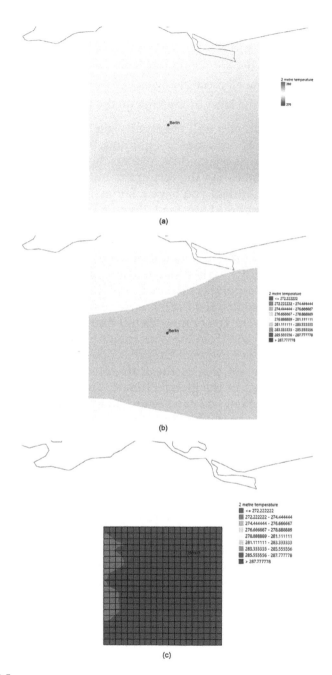

FIGURE 12.5
Berlin (a) overlaid with the Copernicus Atmosphere Monitoring Service air temperature data with the color palette inserted, (b) with the scaling of the color palette changed to discrete with the color palette inserted, and (c) ERA5 air temperature data shown for the same location with the color palette inserted. (Data courtesy of Copernicus Atmosphere Monitoring Service/Copernicus Climate Change Service.)

Go to the menu item Plugins > Manage and Install Plugins. In the Search box type "Crayfish" and one plugin called Crayfish should appear within the right panel, indicating that it is a tool for Mesh layers in QGIS. Click on the Install Plugin button, and once this has finished, click Close.

Go to the menu Mesh > Crayfish > 2D Plot, which puts a graph in the bottom half of the main window, with the Berlin map in the upper half. Currently, the graph is empty, and the various settings in Crayfish need to be assigned. On the list of options at the top of the graph, select the following:

- Layer: CAMS Berlin – although this should be preselected.
- Plot: Time series.
- Group: 2 m temperature for this example, but all the other layers are options.
- From Map: Point. Even if this is selected, still use the downward-facing triangle to bring up the options, and select Point, as this is what activates the plug-in. If you now go onto the map of Berlin in the upper half of the panel, you'll notice the cursor is now a + symbol enabling you to select a point on the map.

If you move the cursor across the map of Berlin, you will be able to see the time series of the 2 m temperature data set from January 01, 2018, on the left side, through to December 31, 2021; similar to Figure 12.6a. If you click the left mouse button, you'll select a specific point, shown as an X on the map. To go back to being able to browse the image, use the downward-facing triangle on From Map and select Point. The graph shows the temperature in Kelvin on the y-axis, but the x-axis is a bit more complicated as it counts hours since January 01,1900, at 00:00:00 and plots these along the bottom as ten to the power of six; yes, we know this is confusing but the upshot is that the left of the graph starts at January 01, 2018, and goes to the right side with December 31, 2021, and for the purpose of the practical don't worry too much about the precise time as it is the trends that you are looking for.

The first point you'll note is the seasonal variations between summer and winter, with the higher temperatures in the summer months, and the four annual cycles evident. However, the graph does not change a lot if you move around the map, and this is because the CAMS data are of a coarse spatial resolution. If you go to the Layer properties for the CAMS Berlin Layer and go to the Symbology option, select the fourth tab, which looks like a grid. Click on the empty checkbox next to Native Mesh Rendering, which will add a tick into the checkbox, and click OK. The CAMS area over Berlin will now be overlaid with a grid showing the data set's spatial resolution, as shown in Figure 12.6b.

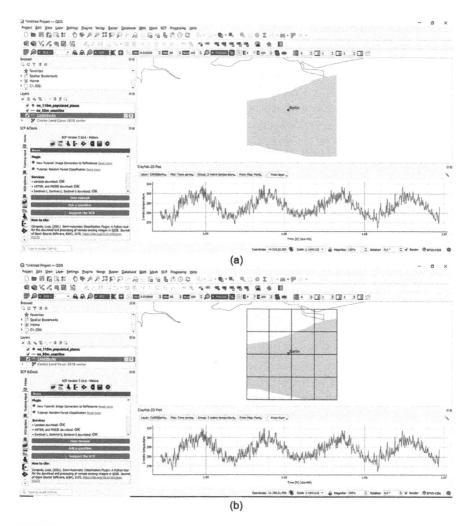

FIGURE 12.6
Copernicus Atmosphere Monitoring Service time series plotted (a) using Crayfish, and (b) with the grid overlaid on the image view. (Data courtesy of Copernicus Atmosphere Monitoring Service.)

You can now investigate the different data set layers, by using the downward-facing triangle by Group, and select one of the other data sets. The data sets show as follows:

- GEMS total column ozone: Like the air temperature data, the ozone data also show four distinct annual cycles, offset from the temperature data; peaking in spring versus summer.

- Total column nitrogen dioxide: This variable also has a strong annual cycle, peaking in winter.
- Particulate matter for $d < 1\,\mu m$ (PM1) and $d < 10\,\mu m$ (PM10): These variables have a much less distinct annual pattern, although it is there, as the concentration of particles in the atmosphere is driven by factors such as wind speed that don't have a distinct annual cycle.

12.6 Summary

This final practical chapter has focused on the remote sensing of the atmosphere, a signal removed as an error source when we try to sense what's on land or in water. We have covered both the short-term and longer-term impacts of humans on what's in the atmosphere and how this affects us, which means we need to track and understand the concentrations of both the gases and particles present. The practical focused on how you can access these data so you can start to explore these data for yourselves.

Overall, this chapter should provide a good starting point, but this book can only give an introduction to a subject that is the focus of many people's careers. We have provided links to online resources, and there are academic textbooks such as Kumar Singh and Tiwari (2022).

12.7 Online Resources

- CAMS: https://atmosphere.copernicus.eu/
- Copernicus Climate Change Service (C3S) Climate Data Store: https://cds.climate.copernicus.eu/
- Copernicus Open Access Hub: https://scihub.copernicus.eu/
- Copernicus Sentinel-5 TROPOMI Science Website: http://www.tropomi.eu/
- CORINE land cover data https://land.copernicus.eu/pan-european/corine-land-cover
- EO Dashboard by NASA, ESA, and JAXA: https://eodashboard.org/
- European Forest Fire Information System https://effis.jrc.ec.europa.eu/

- International Energy Agency's "Methane Tracker" (https://www.iea.org/articles/global-methane-emissions-from-oil-and-gas
- L1 and Atmosphere Archive and Distribution System (LAADS) website http://ladsweb.modaps.eosdis.nasa.gov
- NASA Fire Information for Resource Management System https://firms.modaps.eosdis.nasa.gov/map/
- NASA's OCO-2 and OCO-3 https://www.nasa.gov/mission_pages/oco2, https://www.nasa.gov/mission_pages/oco3
- Natural Earth website https://www.naturalearthdata.com/

12.8 Key Terms

- Aerosols: Particles in the air, which can be classed as pollution.
- Air pollution: Gases or particles in the air that can harm the health of humans, animals, and plants, plus damage buildings.
- Climate change: Refers to long-term shifts in temperature and weather patterns that may be natural, such as through variations in the solar cycle, or due to human activities that have become the main driver.
- GHGs: A GHG absorbs and emits radiation within the TIR range, causing the greenhouse effect.
- Ozone layer: A thin layer in the stratosphere that absorbs almost all the ultraviolet solar radiation, which can be harmful by damaging the DNA of plants and animals.
- Stratosphere: It is about 35 km thick and is located above the troposphere, containing the ozone layer; sometimes called the upper atmosphere. With no storms or turbulence to mix the air, the cold heavier air is at the bottom, and warmer lighter air is at the top.
- Troposphere: The lowest layer of the atmosphere extends from the ground to about 10 km and contains the air we breathe alongside the weather that influences our everyday lives.

References

Bilal, M., J. E. Nichol and P. W. Chan. 2014. Validation and accuracy assessment of a Simplified Aerosol Retrieval Algorithm (SARA) over Beijing under low and high aerosol loadings and dust storms. *Remote Sens Environ* 153:50–60.

Chrobak, U. 2021. The world's forgotten greenhouse gas. Available at https://www.bbc.com/future/article/20210603-nitrous-oxide-the-worlds-forgotten-greenhouse-gas (accessed August 25, 2022).

Dasgupta, S. S. 2020. Air pollution drops in India courtesy COVID-19. Available at https://www.geospatialworld.net/blogs/air-pollution-drops-in-india-courtesy-covid-19/ (accessed August 25, 2022).

ESA. 2021. Monitoring methane emissions from gas pipelines. Available at https://www.esa.int/Applications/Observing_the_Earth/Copernicus/Sentinel-5P/Monitoring_methane_emissions_from_gas_pipelines (accessed August 25, 2022).

Guardian. 2020. China plans rapid expansion of 'weather modification' efforts. Available at https://www.theguardian.com/world/2020/dec/03/china-vows-to-boost-weather-modification-capabilities (accessed August 25, 2022).

Hsu, N. C., S. C. Tsay, M. D. King and J. R. Herman. 2004. Aerosol properties over bright-reflecting source regions. *IEEE Trans Geosci Remote Sens* 42:557–569.

IPCC. 2022. Sixth Assessment Report, Work Group 1: The physical science basis. Available at https://www.ipcc.ch/report/ar6/wg1/ (accessed August 25, 2022).

Kumar Singh, A. and S. Tiwari 2022. *Atmospheric Remote Sensing: Principles and Applications (Earth Observation)*. Amsterdam: Elsevier.

Levy, R. C., L. A. Remer, S. Mattoo, E. Vermote and Y. J. Kaufman. 2007. Second-generation operational algorithm: Retrieval of aerosol properties over land from inversion of Moderate resolution imaging spectroradiometer spectral reflectance. *J Geophys Res Atmos* 112:D13211. http://dx.doi.org/10.1029/2006JD007811.

Logan, T., B. Xi, X. Dong et al. 2010. A study of Asian dust plumes using satellite, surface, and aircraft measurements during the INTEX-B field experiment. *J Geophys Res Atmos* 115:D00K25. http://dx.doi.org/10.1029/2010JD014134.

Lue, Y. L., L. Y. Liu, X. Hu et al. 2010. Characteristics and provenance of dustfall during an unusual floating dust event. *Atmos Environ* 44:3477–3484.

Matsunaga, T. and S. Maksyutov. eds. 2018. *A Guidebook on the Use of Satellite Greenhouse Gases Observation Data to Evaluate and Improve Greenhouse Gas Emission Inventories*, 129. Japan: Satellite Observation Center, National Institute for Environmental Studies.

McMahon, J. 2020. New data show air pollution drop around 50 percent in some cities during coronavirus lockdown. Available at https://www.forbes.com/sites/jeffmcmahon/2020/04/16/air-pollution-drop-surpasses-50-percent-in-some-cities-during-coronavirus-lockdown/ (accessed August 25, 2022).

NASA, 2017. NASA pinpoints cause of earth's recent record carbon dioxide spike. Available at NASA pinpoints cause of earth's recent record carbon dioxide spike (accessed August 25, 2022).

Platnick, S., M. King, and P. Hubanks. 2015. MODIS atmosphere L3 daily product. NASA MODIS Adaptive Processing System, Goddard Space Flight Center, USA. Available at http://dx.doi.org/10.5067/MODIS/MOD08_D3.061

Taylor, T. E., A. Eldering, A. Merrelli et al. 2020. OCO-3 early mission operations and initial (vEarly) XCO_2 and SIF retrievals. *Remote Sens Environ* 251. Available at https://doi.org/10.1016/j.rse.2020.112032

WHO. 2014. 7 million premature deaths annually linked to air pollution. Available at
    https://www.who.int/news/item/25-03-2014-7-million-premature-deaths-
    annually-linked-to-air-pollution (accessed August 25, 2022).
WHO. 2021. WHO global air quality guidelines. Available at https://www.who.
    int/news-room/questions-and-answers/item/who-global-air-quality-
    guidelines (accessed August 25, 2022).

13

Where to Next?

This chapter will focus on the future of remote sensing as a discipline and your personal development within it. We'll start with a review of near-future developments in satellites, sensors, and processing of data and then move on to applications and options for you to continue to learn about remote sensing.

13.1 Developments in Satellite Hardware

This section discusses the developments that will increase the type and quantity of remote sensing data available.

13.1.1 Instruments

Throughout remote sensing's history, as technology has developed so too has the capacity and functionality of sensors and instruments; this is expected to continue. These developments will include the following:

- **Increasing spectral wavebands**: The number of optical spectral wavebands included on instruments has steadily increased; for example, Landsat has more than doubled its number during its mission history. However, the next Landsat mission, currently known as Landsat NEXT, is expected to double those on Landsat-9 with the goal more than to have 25 spectral wavebands on the satellite that is planned for launch in 2029/30.
- **Expansion of the Copernicus Program**: Alongside maintaining the data continuity of the existing Sentinel satellites, the European Union aims to expand the current Copernicus capabilities and has several further missions planned including:
  - Copernicus Hyperspectral Imaging Mission for the Environment (CHIME) to carry a visible to shortwave infrared spectrometer to provide routine hyperspectral observations.

DOI: 10.1201/9781003272274-13

- Copernicus Imaging Microwave Radiometer (CIMR) to carry a multi-frequency microwave radiometer to provide observations of sea-surface temperature, sea-ice concentration, and sea-surface salinity.

- Copernicus Anthropogenic Carbon Dioxide Monitoring (CO2M) to carry a near-infrared and shortwave-infrared spectrometer to measure atmospheric carbon dioxide produced by human activity.

- Copernicus Polar Ice and Snow Topography Altimeter (CRISTAL) to carry a dual-frequency radar altimeter and microwave radiometer to measure sea-ice thickness and overlying snow depth.

- Copernicus Land Surface Temperature Monitoring (LSTM) to carry a high spatial-temporal resolution thermal infrared sensor to provide observations of land-surface temperature.

- ROSE-L: Copernicus L-band Synthetic Aperture Radar (SAR) to carry an L-band SAR to work in conjunction with the C-band on Sentinel-1.

- SAR: As discussed in Chapter 3, the number of SAR satellites in orbit is increasing rapidly, and they are seen as a key part of the overall solution of exploiting satellite data as they can see through clouds and smoke and so can capture data when optical satellites cannot. In addition to the SAR constellations mentioned in Chapter 3, there are also constellations such as Umbra Lab's X-band SAR constellation, the Chinese Gaofen-3 C-band SAR satellites and CHORUS from the Canadian company MDA. It is expected that the number of SAR missions will continue to increase, for example, as noted above, with ROSE-L, and a new Chinese commercially run SAR constellation known as 36 Tiangan expects to begin launches in 2022, while the existing players such as Capella Space and ICEYE will continue to develop their capabilities.

- Hyperspectral data: Hyperspectral satellites operated from a lower base than SAR satellites. However, the increasing sophistication of computing power means that effectively processing the hundreds of wavebands of hyperspectral is becoming easier. This increased capability has led to several launches in recent years, such as the following:
  - PRIMSA (PRecursore Iperspettrale della Missione Applicativa) satellite was launched in March 2019, by Italy's national space agency.
  - German Environmental Mapping and Analysis Program (EnMAP) satellite went into orbit in April 2022.

- GHGSat-3, GHGSat-4 and GHGSat-5 are Greenhouse Gas Satellite Demonstrators offering hyperspectral SWIR imaging targeted at monitoring greenhouse gases, operated by GHGSat Inc and launched in June 2022.
- Commercial company Pixxel launched the first of its high-resolution hyperspectral microsatellites in April 2022; the company aims to have a constellation of six satellites that could image the planet every 2 days.
- China's pair of Goafen-5 hyperspectral satellites launched in May 2018 and September 2021, respectively.

Again, it is likely that there will be more and more hyperspectral imagers launched in the coming years, with CHIME noted above, plus the National Aeronautics and Space Administration (NASA) Surface Biology and Geology mission, which aims to launch in 2027.

13.1.2 Satellite Developments

13.1.2.1 Smaller and Smaller Satellites

One of the biggest trends of the last decade in Earth Observation (EO), and the broader satellite industry, has been the emergence of smaller satellites. Historically satellites were large, for example, Envisat (carrying Medium Resolution Imaging Spectrometer [MERIS]) weighed approximately 8,200 kg and was the size of a double-decker bus, which meant it was expensive to develop, build and launch. However, there has been a focus on developing smaller satellites as they are much cheaper and quicker to design, build, and launch. This new approach has opened up the satellite industry to new organizations, rather than just the traditional space agencies and large companies. The most common types of smaller satellites are as follows:

- **CubeSats**: CubeSats, also called nanosatellites, are so named as they are composed of several 0.1 m cubes, with each cube having a mass of 1.33 kg. Several cubes can be combined to create a larger satellite; for example, Planet Labs Flock 1 constellation satellites are comprised of three cubes (referred to as 3U). CubeSats have also opened space development and exploration to educational establishments, with some helping students to build, design, and launch their own satellite. These CubeSats also offer fantastic options to build prototypes, test new sensors, or undertake specific short-lived measurements.
- **Picosatellites**: These are tiny satellites, only a few cubic centimeters in size and between 100 g and 1 kg in weight, and are about the

size of one-quarter of a CubeSat. These types of satellites are often launched as small constellations and work together as though there was a single larger satellite; they are known as swarms. For example, the commercial company Swarm Technologies has a constellation of picosatellites known as SpaceBEES that are used for communication applications.

These small satellites have a much shorter mission life than larger satellites, and mostly burn up during reentry to the Earth's atmosphere at the end of their life.

The growth of CubeSats has also led to the development of a new market in the satellite launch industry, with the Indian Space Research Organization (India's space agency) and the commercial companies SpaceX and Rocket Labs. These organizations offer so-called 'ride share' launches, where many satellites are sent into orbit together on a single rocket, and they all share the launch cost making it cheaper. In January 2021, Space X set the world record, at the time of writing, when they put 143 satellites of various shapes and sizes into orbit on a single launch.

13.1.2.2 Constellations

The second big trend in recent years has been the development of satellite constellations. The generation of new space companies has led to this, but traditional space agencies have also been brought along, for example, the European Space Agency (ESA) Copernicus Program has twin satellites for each of its core missions, such as Sentinel-1A and 1B, or Sentinel-2A and 2B with further pairs (C and D) prepared for launch so they can be used to create a continuous time series of data. This has the benefit of not only reducing temporal resolution by allowing data to be collected more frequently but also providing a level of resilience for data provision that is critical for commercial companies building and selling services on the back of satellite data; the importance of this is demonstrated by the loss of Sentinel-1B as described in Section 6.4. It is not just ESA, for example, the Canadian Space Agency's RADARSAT is a multiple satellite constellation, and the next section describes some of China's constellations.

Despite these examples, the new space companies pioneered the use of CubeSats constellations. In EO terms, Planet Labs is best known with over 200 satellites in space providing daily optical images. However, they are certainly not alone, with companies such as BlackSky and Satellogic offering optical constellations, and SatelliteVu looking to launch a thermally focused constellation. In addition, in the previous section we highlighted SAR constellations. It is challenging to keep up with the number of companies and their plans, but the NewSpace Index (https://www.news-pace.im/) provides a comprehensive list alongside ESA's list of missions (https://earth.esa.int/eogateway/missions) that includes their own and those whom they provide data access.

However, it's outside EO, in the space-based internet broadband application where constellations really exploded. The most famous of which is SpaceX who, at the time of writing, have over 2,000 of their Starlink satellites in orbit; although this pales into insignificance with their plans to launch 12,000 satellites in the coming few years, and eventually, they hope to have a constellation of over 40,000 satellites. To put that into context, including Starlink's current satellites, the world has only launched just over 13,000 objects into space ever.

It's not just Starlink, other space-based internet broadband suppliers include OneWeb on the way to its planned 648 constellations, and Jeff Bezos is looking to launch several thousand satellites with his Project Kuiper. All of this means that space will become very congested in the coming years, which could cause EO issues. For example, in 2019, ESA had to undertake a "collision avoidance maneuver" for its Aeolus EO wind satellite to prevent a potential collision with a SpaceX Starlink satellite. There are also complaints that these satellites are ruining the view of the stars for astronomers; it is possible to see Starlink satellites passing across the night sky with the naked eye.

13.1.2.3 China

Over the last decade, China has significantly developed its EO capabilities and now, arguably, is the world's EO data collection powerhouse. One of the most ambitious projects was the China High-resolution Earth Observation System (CHEOS), organized by China National Space Administration, which was initiated in 2010. Since that time, they have launched a range of optical, SAR, and hyperspectral GaoFen satellites; GaoFen means high-resolution in Chinese. CHEOS aims to provide the country with all-weather all-day global coverage. At the time of writing, there are believed to be at least 30 satellites within the system. However, CHEOS may not only involve satellites with potential future developments to include airborne and High-Altitude Pseudo-Satellite (see Section 13.1.2.5) components.

This extension of remote sensing capabilities beyond satellites is supported by the Chinese Aeronautic Remote Sensing System, which became operational in 2021. Developed by the Aerospace Information Research Institute of the Chinese Academy of Sciences, it consists of two medium-sized crewed aircraft with more than ten remote sensing instruments on board allowing the collection of data from multispectral, multi-temporal, multi-polarization, and multi-angle viewpoints.

In addition to CHEOS, other constellations have been launched, such as the Yaogan satellite constellation of over 50 satellites. Little detail is known about the specifications of these satellites as they are believed to be focused on military purposes.

Data access by anyone is becoming more possible with Chinese commercial firms actively involved in EO; for example, the Chang Guang Satellite Technology Company has designed and launched a constellation of over 30 Jilin-1 optical imaging satellites with this data sold commercially worldwide. Secondly, the Chinese National Space Administration announced in April 2022 that they had established a satellite data center described as being a hub for the international exchange of satellite data, with particular relevance to the countries of Brazil, Russia, India, China, and South Africa. These five signed an agreement in 2021 to enhance the sharing of remote sensing data.

It is also worth noting that in 2021 China completed its Beidou Navigation Satellite System to rival the American GPS and European Galileo systems. In 2022, they also completed their global EO remote sensing calibration benchmark network, providing an efficient calibration infrastructure and scientific test platform in accordance with international standards.

13.1.2.4 Democratization of Space

The introduction of lower-cost satellites and launches has also led to a democratization of space and, as we've already talked about, the increase in commercial providers and educational organizations that have become involved in EO. However, it is also true that many more countries now have their own access to both satellites, especially EO satellites. Some of these are geostationary satellites focused on their own countries with the data used for various applications.

According to the active satellite database maintained by the Union of Concern Scientists, in 2021, there were over 60 individual countries that had control of at least one EO satellite. The number of countries with this sort of access continues to grow, which is important as it means those countries are not reliant on getting data from other organizations, bodies, or countries. It gives them control and independence over data and how to use it. However, it's also vital that data are shared for scientific and disaster-related purposes, as discussed further in Section 13.2.4.

13.1.2.5 High-Altitude Pseudo-Satellite/High-Altitude Platform Station

In Chapter 1, we acknowledged that there are other ways to undertake remote sensing besides satellites. In this book, we have focused on satellites. Still, looking forward, it is worth noting non-satellite developments, particularly High-Altitude Pseudo-Satellite/High-Altitude Platform Station (HAPS), also sometimes referred to as atmospheric satellites.

HAPS is a platform operating at high altitudes, around 20 km above the Earth, for extended periods effectively providing similar services to satellites. HAPS can be airships, planes, or balloons – that float, or fly, in the stratosphere. ESA has had a HAPS program since 2010, launching several prototypes. While in the United States, the Defense Advanced Research Projects Agency, has tested a balloon carrying the Stratospheric Optical Autocovariance Wind LiDAR that floated above the same position on the Earth, similar to a geostationary satellite. Commercial organizations are also exploring this area, with Thales Alenia Space developing the Stratobus designed to carry both radar and optical imaging technologies, allowing for day, night, and all-weather observations.

It will be interesting to see how these new technologies will develop over the coming years and what new potential they offer for EO.

13.1.2.6 Uncrewed Aerial Vehicles

Uncrewed aerial vehicles (UAVs), known in everyday life as drones, were mentioned as a rapidly developing remote sensing component in the previous edition. These airborne vehicles offer a wide range of applications and have the advantages of increased spatial resolution and the ability to fly on request.

They have become a popular and highly used technology for applications that require limited geographic areas to be surveyed, such as agriculture, glaciology (studying glaciers), construction, and urban planning. For applications such as glaciology and construction, it is about the ability to undertake Structure-from-Motion photogrammetry, that is, by having a series of images/video a 3D model can be developed. For agriculture, it is the combination of artificial intelligence (AI) and UAVs that have created an ability to leverage so-called "big data" with crop information extracted from UAV data driving more accurate satellite-based crop status information at large scales (Jung et al., 2021); the use of AI with EO will be discussed further in Section 13.2.5.

However, the need for them to be flown by a person with line-of-sight of them has restricted their usage. Like with autonomous cars, the technology can potentially do more than it is currently allowed for privacy, safety, and legislative reasons. This restriction may well change shortly as their ability to reduce cost and impact drives further change.

13.1.2.7 Sustainability: Space Debris and Carbon Footprint

To date, there have been just over 13,000 objects ever launched into space, with only 30% of these objects having returned to Earth. Of the remaining 9,500 objects that are still in orbit, only around 57% are actually operational; the other 4,000 are space debris, and there is no easy way to remove them. All new satellites going up need to plan on how to safely bring them back to Earth at the end of their mission, but this was not always the case. Therefore, the inactive satellites don't have any fuel to bring them back and are left in space. There are plans, and some prototype missions, to try and capture space debris, slow down their speed or knock them down to lower orbits so they will fall into the Earth's atmosphere and burn up. To date, none of these ideas have provided a clear solution.

Any collision, even from a fleck of paint, can be serious. However, despite the near collision we mentioned above in Section 13.1.2 and many more near-misses than you might think, satellite orbits have to be regularly adjusted as it is not the satellites themselves that are the real danger, but much smaller pieces of debris. It's estimated that there may be 300 million pieces of debris in space, which are traveling at speeds of up to 33,500 miles per hour, equivalent to 558 miles per minute. In 2016, Copernicus Sentinel-1A was struck by a millimeter-size object damaging its solar array. These smaller pieces can be anything from tools lost by astronauts to debris from destroyed satellites; in 2021 Russia conducted an anti-satellite missile test blowing up one of their inactive satellites, and it created 1,500 pieces of new trackable space debris and many more thousand smaller pieces, while in 2007, China destroyed one of its weather satellites creating over 3,000 larger pieces of debris and over 100,000 smaller pieces.

Beyond the potential collision risk, there is a greater potential danger as outlined by Donald Kessler in his 1978 Kessler Syndrome hypothesis, which proposed a scenario where the density of objects in low earth orbit is so great that the debris from a single collision between two objects would set off a cascade of collisions that would stop any further spacecraft from passing through the low Earth orbit area – preventing anything else from safely leaving the Earth. While increasing the number of satellites in orbit significantly, like Starlink is doing, does offer potential new benefits, care must be taken to ensure it does not put the entire industry and human space flight at risk.

Space debris is one aspect of sustainability, another is carbon footprint. While the EO industry stands up and offers solutions to monitoring changes in the climate, it can't ignore the fact that to do this it has to launch satellites, which is not a carbon neutral activity. The importance of the industry improving its environmental contributions is gaining traction. ESA recently set some European targets as part of its Agenda 2025 program. In particular, ESA set a target of reducing greenhouse gas emissions by 46% by 2030 compared with its 2019 baseline, and to continue towards the climate neutrality of Europe by 2030 as per the Paris Agreement objective of keeping temperature increase to within 1.5°C by 2100.

The carbon footprint of the space industry isn't easy to calculate, although some interesting papers have been produced. Wilson et al. (2022)

focused on the Life Cycle Sustainability Assessment methodology using all space missions launched in 2018. The research discovered that in terms of greenhouse gases, 84.90% of the total emissions came from building spacecraft, launcher components, and fuels, including their management, handling, and storage, due primarily to the carbon dioxide released during heat and electricity consumption. This issue will become increasingly important in the coming years.

13.2 Developments in Data Processing

As the transmission technology onboard satellites has improved, it's possible to send increasing amounts of data back from space; data relay satellites also allow this to be increased further. This means the amount of remote sensing data is growing exponentially; for example, the Sentinel-1A mission is producing around 1.7 TB of Level 1 (L1) data per day, and the Sentinel-2A mission produces about 800 GB of raw compressed data every day. All these data need to be processed when they arrive on Earth. Initially, this will involve the calibration and conversion into a format suitable for the users: raw data to Level 0 and then L1. Then the onward processing to Level 2 (L2) and more user-friendly options as discussed in Section 13.2.7.

Therefore, users can access large data sets through the Internet and want to answer their questions. The rest of this section describes some current developments in data processing.

13.2.1 Accessing Online Data Sets

Traditionally, acquiring data has involved downloading them from the Internet as you have done within the practicals. However, there are increasing alternatives to downloading, by accessing, using, and even processing data held by online providers without the need to download anything. This reduces the local storage space and the computer processing power required to undertake image processing. There are several options to consider:

- **Google**: Google Earth is probably the best-known source of high-resolution satellite imagery – although the imagery is acquired by separate satellite and airborne remote sensing organizations. It is entirely free to use and covers the globe, and although the data are not always the most recent, they tell you the year of the image in the bottom corner alongside who acquired it in the acknowledgments. It is excellent if you want to explore the Earth and is easy to use, but it is not downloadable, and you can't go back in time. However, a different version called Google Earth Pro can be downloaded and provides more capabilities. There is also a more scientific version known as Google Earth Engine, an online platform allowing users to process, analyze, and visualize satellite data. It has a wide range of current and historical satellite data, including Sentinel-2, Sentinel-3, and Sentinel-5P; all the Landsat data sets; and many others. It not only offers some simple analysis tools but also provides the option to build computer code on the platform to perform more complex analyses. Google Earth Engine is a free service for scientists, with a more recent paid-for service for those wanting to use it for commercial purposes.

- **Amazon**: Amazon has its own network of ground receiving stations, and they download data directly from satellites and make it available within their cloud computer storage, called Amazon Web Services. As part of this, they provide tools that you can build to process and visualize the data online – you can set up virtual machines (computers) that exist for as long as you want. Available data sets include Sentinel-1 and Sentinel-2, Landsat-8 and Landsat-9, and others. The tools provided are getting better, but satellite data knowledge is still needed to use them effectively. This approach is useful if you don't have much local storage or computer power. However, it is not a free service, and every step has a cost, particularly if you want to download the final results.

- **Microsoft**: The third of the big three computer service providers has the Microsoft Planetary Computer, which offers a catalog of global satellite and environmental data sets together with tools that help users develop applications to analyze, process, and visualize data. The tools include simple online mapping solutions, an application to build more complicated analyses, connections to external data solutions to allow the data to be used, and it can also be connected to packages like QGIS. Their catalog includes Sentinel, Landsat, MODIS, National Oceanic and Atmospheric Administration (NOAA) data, and some Planet Labs data. The data in the system are free to use, but again you will pay for cloud computing.

- **Web services**: In Chapter 12, you used the CORINE data set as a web service, where you connected your local computer to online data sets without downloading them. This approach offers the advantage that you can continue to do your analysis locally, but the input data will not take up hard disk space. There are free-to-use services such as CORINE, but there are also commercial services such as the Sentinel Hub that offers all Sentinel and Landsat data, together with MODIS, ENVISAT, and other commercial data sets. While you can download data freely from Sentinel Hub, using their web services has a monthly cost that depends on how much data you use. The services above from Google, Amazon, and Microsoft also offer web service options.

- **Data and Information Access Services (DIAS)**: These five online platforms were developed to make it easier to access Copernicus data and information and promote its use (https://www.copernicus.eu/en/access-data/dias). They provide access to all Sentinel data, along with many of the Copernicus services, such as the data on the Copernicus Atmosphere Store. Alongside the data, they also host applications and tools to help users process, analyze and visualize data. The five DIASs are run as commercial operations, so there is a cost to using them; this varies between the different DIASs as do the specific services and tools they offer.

13.2.2 Onboard Satellite Data Processing

Onboard satellite data processing is a future trend. Once a satellite acquires data, the satellite performs data processing in space instead of sending it to Earth. This processing might be a simple task such as compressing the data, similar to putting it into a zip file, so it can be downlinked faster. Alternatively, more complex analysis can be done, such as looking at the image's cloud cover; if it's too great, it will not send the data back. Therefore, focusing on sending the most usable data saves time, costs, and storage space.

The ESA-funded ɸ-sat, pronounced PhiSat, program (https://philab.esa.int/explore/the-phi-lab-explore-office/phi-sats-programme/) is focused on AI in space. ɸ-sat-1, launched in August 2020 based on two CubeSats, demonstrated onboard feature detection using hyperspectral EO data and AI. The next ɸ-sat-2 mission will further demonstrate the capabilities of AI technology for EO by being capable of running AI apps onboard via a computer from Open Cosmos.

Another example of how far this technology might go was shown in 2022 when Wuhan University in China, described the Luojani-3 01 satellite it hoped to launch which would have onboard data analysis and processing using AI allowing it to transmit user-friendly data back to earth rather than raw data. With a high-resolution optical camera onboard, it hopes to allow anyone with a smartphone to access near-real-time images and videos from space.

D-Orbit, a company focused on space logistics and transportation services, is also interested in this technology, as they foresee the potential to offer significant benefits for the industry. They completed orbital testing of Nebula, a cloud platform designed to provide distributed high-performance data analytics computing and storage capabilities in space that enables end-users to uplink and run software such as AI apps in a way similar to conventional, terrestrial cloud-based environments.

13.2.3 Integration

Historically, many data sets were held by government agencies, and slowly over recent years, more and more of these have been made available to the broader community; however, this process still has a long way to go to make the majority of data available to everyone.

Data sets have tended to be held in isolation. Yet within the industry, there is a clear recognition that bringing disparate data sets together can create a greater combined value. In simple terms, integrating optical and SAR data would allow images to be collected irrespective of the weather conditions. This approach is something Planet Labs offers, where it merges SAR data from Sentinel-1 with the optical images from its fleet. However, this impact goes far beyond the images, with data from space agency satellites being used to calibrate commercial satellites.

Within Europe, the INSPIRE directive provides a strong impetus, emphasizing the sharing of environmental spatial information among public sector organizations to facilitate public access better. This concept is gathering momentum in the wider remote sensing community. The Open Geospatial Forum (OGC) has also undertaken several projects looking at how to integrate data, such as the Disaster Pilot projects, which have been looking to put in place steps to make data sets, including satellite data, available more rapidly to first responders when disasters occur to enable them to make decisions quickly.

13.2.4 Machine Learning and Artificial Intelligence

Machine learning (ML), a subset of AI, can be applied across most applications and ranges from image processing to high-level data understanding and knowledge discovery. One of the most common applications is determining the relationship between the satellite, multiple satellite signals, and features of interest. In the practicals, we have already worked on such as problem in Chapter 9 where we applied a supervised classification. Using an ML technique such as Random Forests or an artificial neural network, often shortened to a neural network, requires following a similar procedure. Random Forest is based on the iterative and random creation of decision trees, that is, a set of rules and conditions that define a class, while neural networks are inspired by the human brain's neurons and our ability to learn.

The first crucial step is to create a training data set that captures the variability within the satellite data and the feature of interest. As we mentioned in Chapter 9, this can involve digitizing tens of thousands of locations so that it is comprehensive. Alternatively, it is also possible to go to sites such as the Radiant Earth Foundation ML Hub (https://www. radiant.earth/mlhub/) and gain access to training data someone else has digitized. Then, the user needs to select an appropriate technique that is applied and validated/tested by splitting the training data up into independent subsets used for training, validation, and increasingly independent testing.

D'Amour et al. (2022) investigated that ML models often exhibit unexpectedly poor behavior when deployed in real-world domains. This drop in performance was identified as being caused by underspecification, where observed effects can have many possible causes. Therefore, as training/validation can be part of the model definition process, having an independent subset not previously used can further understand future behavior. The real strength of these models is that once they have been trained, you can apply them to new data and gain insights.

In SNAP, you have access to classifiers such as Random Forest and neural networks under the Raster > Classification > Supervised Classification menu item. In addition, the Semi-Automatic Classification Plugin we used in QGIS has a Random Forest option that uses this functionality from SNAP. An example of the Random Forest plugin in SNAP being applied is Page et al. (2020), where tire and plastic waste in Scotland were classified using Sentinel-1 and Sentinel-2 data.

13.2.5 Open Source and Open Science

As demonstrated in this book, many software creators provide free and open access to what they've developed. This doesn't mean paid-for software is no longer needed, but it provides users with a choice; being part of an open-source community should involve both give and take, with users volunteering their time and providing supporting funds. QGIS is an excellent example of open-source software as it has developed and improved due to this community effort and support, and is an accepted tool across a range of organizations.

Linked to open-source software developments are changes in preferred programming languages, with Python currently being popular alongside statistics programming languages such as R. These open-source development languages mean that both the code and final software are free to both download and amend. An example is the Remote Sensing and GIS Software Library (RSGISLib, http://www.rsgislib.org/).

Growing in parallel with open-source software is open science (https://openscience.org/what-exactly-is-open-science/), which has four fundamental goals:

- Transparency in experimental methodology, observation, and collection of data.
- Public availability and reusability of scientific data.
- Public accessibility and transparency of scientific communication.
- Using web-based tools to facilitate scientific collaboration.

For example, NASA has made a long-term commitment to build an inclusive open science community over the next decade. It's also key to ESA's Digital Agenda for Space, and in the support, they offer data in collaborative environments to further facilitate the exchange of ideas among scientists. In addition, by adopting open science principles, scientists can advance climate change research and accelerate efforts to mitigate impacts; especially for highly vulnerable developing regions where research capacity may be limited.

13.2.6 Data Standards

There is an increasing movement towards data standards within the remote sensing industry. Historically, satellite data had limited commonality between satellites as each data set had its own structure and content. It has been recognized that this approach makes it more difficult for users to work with the data, as they need to understand multiple data sets to undertake analysis. Several standardization approaches have started to take hold to improve the uptake and use of data, which is being driven by cloud-based data processing, as described in Section 13.2.1. We need consistent ways to describe and handle satellite data, and this is just a brief introduction to some of the ongoing work:

- Analysis ready data: As an alternative to the raw data or the various levels of data we've referred to in this book, a recent trend is the development of Analysis Ready Data (ARD) data sets. These satellite data sets have been processed to a minimum set of requirements and organized into a form that allows immediate analysis and integration with other data sets with minimum user effort. CEOS is one of the organizations leading on agreed standards for ARD data (http://www.ceos.org/ard/).
- Data cubes: A data cube simply describes a multi-dimensional array of values where the data contains one or more spatial or temporal dimensions. For the EO data cube, the ARD data sets

are critical to allow the data to be stacked into a cube both horizontally and vertically, that is, geographically, and temporally. Data from multiple satellites can be stored in one data cube and analyzed altogether, removing from the user the need to reproject data into consistent map projections as we undertook in Chapter 11.

- Data storage and access standards: OGC is one of several organizations jointly developing standards to make location information and services Findable, Accessible, Interoperable, and Reusable (FAIR). Software developers can then use these to build solutions that can more easily share data using consistent structures and languages, making it simpler for users to use multiple data sets. (https://www.ogc.org/standards). An example is SpatioTemporal Asset Catalogs (STAC) which catalogs spatiotemporal asset representing information captured in a certain space and time; in essence, every satellite data file is a spatiotemporal asset. The aim is that all providers of spatiotemporal assets will catalog their data, in the same way, to make it easier for users to index, discover and extract data sets. We saw that Landsat Collection 2 data are now provided with a STAC metadata file, and if you access EO data stored on cloud-based platforms such as AWS there will be STAC catalogs you can use to both search and view potentially interesting data sets.

13.3 Developments in Applications

The ways in which remote sensing is used are changing, and the applications for which it can be used are expanding. Some of the current developments in applications are discussed in the succeeding sections.

13.3.1 Citizen Science

Citizen science is the collaboration between scientists and the wider general public and is increasingly part of open science. It is based on the principle of collective intelligence, the idea that, under the right circumstances, collections of individuals are smarter than even the smartest individuals in the group (Surowiecki, 2004). It underpins much of the enthusiasm for crowd-sourced data, where members of the public can collect larger quantities of data over a wider geographical coverage than would be impossible by researchers alone.

There are an increasing number of citizen science projects using satellite data that you can get involved with. This might be to validate remotely sensed data with in situ measurements or to undertake object recognition where you confirm certain features in an image that helps train ML systems. These processes are very time-consuming, and it would be costly if volunteers were not used. For citizens involved, these projects are great fun to participate in, and for the scientists, the volunteered geographic information provides an opportunistic source of information with enthusiastic citizens providing viewpoints that wouldn't otherwise be available.

Some examples of these types of projects include:

- NASA's Citizen Science: These are projects available for anyone in the world to participate in, focusing on issues both on Earth and in outer space. (https://science.nasa.gov/citizenscience).
- Citizen Science Earth Observation Lab (CSEOL) is an ESA-funded initiative to encourage projects using data to empower citizens, improve decision-making and examine issues using satellite data and citizen science. (https://cseol.eu/)
- Open Street Map: An example of a web-based collaborative mapping system (https://www.openstreetmap.org).

13.3.2 Climate Quality Data Sets

Climate studies require consistently calibrated observations with adequate temporal and spatial sampling over a long period. As the repeat cycle of any single satellite mission will limit the creation of such a data set, the Global Climate Observing System Implementation Plan (GCOS, 2010) set out a series of actions concerning Essential Climate Variables (ECVs), as discussed in Chapter 9, which the United Nations Framework Convention on Climate Change required nations to respond to. In response, ESA Climate Change Initiative (CCI) projects (https://climate.esa.int/en/) were funded to integrate multimission data to create climate quality data sets, combining L3 and L4 products. There are individual project websites to access the data, but ESA has developed a single portal and online software tool to support the use of multiple data sets. Other space-related organizations are also meeting this requirement; for example, NOAAs are building the Global Land and Marine Observations Database to increase understanding of climatic variability, extreme past weather, and climate events to help regions, countries, and stakeholders make better decisions.

A different type of climate data set is the Fundamental Climate Data Record (FCDR), a long-term data record of calibrated and quality-controlled data designed to generate standard products that are accurate and stable enough for measuring climate variability and climate change. This data record differs from the ECVs in that instead of merging the derived product, it is the underlying data sets that are brought together, and then a consistent product is generated from the FCDR. This approach requires detailed work on understanding how different instruments collect data, and the errors and anomalies that affect each instrument need to be accounted for.

13.3.3 Repurposing

Remote sensing techniques tend to be developed out of the area of interest of an individual researcher and hence applied to one particular field. Repurposing is the extension of a technique, or data set, into an area it either hasn't been previously used in or potentially was not designed for.

It requires additional research but offers the potential to extend the use of existing systems to different applications. Examples of repurposing include the following:

- Satellite: Several satellite instruments have been put to a secondary use once in orbit; for example, Cryosat-2 is being used to map much more than sea ice.
- Technique: Lidars have also been proposed for the ocean, as they offer the prospect of vertically resolved measurements, while ocean color provides a surface layer integrated value. The detection of an oceanic signal has already been demonstrated by the atmospheric-focused, Cloud-Aerosol Lidar and Infrared Pathfinder Satellite Observation satellite mission (Behrenfeld et al., 2013).
- Location: In April 2015, NASA, NOAA, the United States Geological Survey (USGS), and the US Environmental Protection Agency announced a $3.6 million partnership to research the application of ocean color techniques with inland waters, to develop an early warning system for harmful freshwater algae blooms.

13.4 Developing Your Knowledge Further

This book has described the basic concepts, applications, and techniques of remote sensing and image processing, including understanding processing and handling data within a GIS. We've tried to give you a broad introduction to the subject, and therefore there's still much more you can learn. We've only scratched the surface of many application areas, and there is much more to learn about what we've described. There are also many remote sensing applications we've not had space to include in this book. We've also kept the language as understandable as possible and equations to a minimum, simplifying what are often complex techniques.

If you're interested in learning more about remote sensing, several textbooks and articles are given in Section 13.4.1, or online resources have been given throughout the book. Don't forget that there's also a website accompanying this book (https://www.playingwithrsdata.com/), which will be kept updated with various remote sensing resources.

There are also several Massive Open Online Courses, known as MOOCs, which cover various aspects of using, processing, and visualizing satellite data. Some courses are free to undertake, while others do have small fees. Examples of well-known MOOC providers include Future Learn (https://www.futurelearn.com/) and Coursera (https://www.coursera.org/), although there are also other providers and occasional specific satellite courses run by space agencies.

Another exciting way of developing your skills would be to participate in a Hackathon. These social events take place over a couple of days where a team of people is brought together to improve on or build a new software program or app on the theme of the Hackathon. You don't have to be a computer coder to participate in the events, as the teams tend to have various skills. For example, NASA's Space Apps Challenge runs global hackathons at various locations and online (https://www.spaceappschallenge.org/). Although, again, there are plenty of other hackathons that get advertised.

We hope this book has whetted your appetite for remote sensing, and the practical exercises have shown you the possibilities. Enjoy playing and experimenting with the techniques to undertake your own remote sensing. Investigate issues and questions that excite and interest you; who knows what you might discover? The next step is up to you.

13.4.1 Examples of Further Reading

- Barale and Gade (2008): Remote sensing of European Seas.
- Green et al. (2000) and Mumby et al. (1999): Coastal management of tropical resources.
- Kumar Singh and Tiwari (2022): Atmospheric Remote Sensing: Principles and Applications
- Lakshmi (2015): Remote sensing of the terrestrial water cycle.
- Manakos and Braun (2014): Land Use and Land Cover remote sensing.
- Miller et al. (2005): Coastal remote sensing for aquatic environments in general.

- Njoku (2014) and Warner et al. (2009): General remote sensing books.
- Robinson (2004): Marine remote sensing.
- Russ (1992): Image processing, with Chapter 2 explaining the correction of image defects and Chapter 3 focusing on image enhancement.
- Weng (2014): Global urban monitoring.
- Woodhouse (2006): Microwave remote sensing.

13.5 Summary

We've reached the end of our journey into remote sensing. Our aim was to interest, intrigue, and excite you about the possibilities this industry offers in terms of understanding what is happening on and to our planet.

You've gained a toolkit of theory, software, and techniques that you can now use to undertake your remote sensing in whatever field you're particularly interested in. We hope you found the book, and the practical exercises within it, enjoyable and interesting. Still, more than that, we hope you carry on exploring and developing your knowledge of remote sensing.

Don't forget to visit the online learning resources accompanying this book (https://www.playingwithrsdata.com/), or our Pixalytics website (https://www.pixalytics.com/), to see what's currently happening in the world of remote sensing. We'd love to hear feedback on the book and your remote sensing experience.

13.6 Online Resources

- Associated learning resource website: https://playingwithrsdata. com/
- CSEOL: https://cseol.eu/
- Coursera MOOC Provider: https://www.coursera.org/
- Data and Information Access Services https://www.copernicus. eu/en/access-data/dias
- ESA climate science page: https://climate.esa.int/en/

- ESA's first massive open online course (MOOC) titled "Monitoring Climate from Space": https://www.futurelearn.com/courses/climate-from-space
- Future Learn MOOC Provider https://www.futurelearn.com/
- NASA Citizen Science Projects: https://science.nasa.gov/citizenscience
- NASA Earth Observatory: http://earthobservatory.nasa.gov/
- NASA's Space Apps Challenge: https://www.spaceappschallenge.org/
- Open Geospatial Forum Standards: https://www.ogc.org/standards
- Open Street Map: https://www.openstreetmap.org
- EO College: https://eo-college.org/welcome
- Remote Sensing and GIS Software Library (RSGISLib): http://www.rsgislib.org/

References

Barale, V. and M. Gade. 2008. *Remote Sensing of European Seas*. New York: Springer-Verlag.

Behrenfeld, M. J., Y. Hu, C. A. Hostetler et al. 2013. Space-based lidar measurements of global ocean carbon stocks. *Geophys Res Lett* 40:4355–4360.

D'Amour, A., K. Heller, D. Moldovan et al. 2022. Underspecification presents challenges for credibility in modern machine learning. Preprint at https://doi.org/10.48550/arXiv.2011.03395

GCOS. 2010. Implementation plan for the global observing system for climate in support of the UNFCCC (2010 Update). Available at http://www.wmo.int/pages/prog/gcos/Publications/gcos-138.pdf (accessed April 17, 2015).

Green, E. P., P. J. Mumby, A. J. Edwards, J. Alasdair, C. D. Clark. 2000. Remote sensing handbook for tropical coastal management. Available at https://unesdoc.unesco.org/ark:/48223/pf0000119752 (accessed August 25, 2022).

Jung, J., M. Maeda, A. Chang et al. 2021. The potential of remote sensing and artificial intelligence as tools to improve the resilience of agriculture production systems. *Curr Opin Biotechnol* 7:15–22. https://doi.org/10.1016/j.copbio.2020.09.003.

Kumar Singh, A. and S. Tiwari. 2022. *Atmospheric Remote Sensing: Principles and Applications (Earth Observation)*. Amsterdam: Elsevier.

Lakshmi, V. 2015. *Remote Sensing of the Terrestrial Water Cycle*. Hoboken, NJ: John Wiley & Sons Inc.

Manakos, I. and M. Braun. 2014. *Land Use and Land Cover Mapping in Europe: Practices and Trends*. Berlin, Heidelberg: Springer-Verlag.

Miller, R. L., C. E. Del Castillo and B. A. McKee. 2005. *Remote Sensing of Coastal Aquatic Environments: Technologies, Techniques and Applications*. New York: Springer-Verlag.

Mumby, P. J., E. P. Green, A. J. Edwards and C. D. Clark. 1999. The cost-effectiveness of remote sensing for tropical coastal resources assessment and management. *J Environ Manage* 55(3):157–166.

Njoku, E. G. 2014. *Encyclopedia of Remote Sensing*. New York: Springer-Verlag.

Page, R., S. Lavender, D. Thomas et al. 2020. Identification of tyre and plastic waste from combined copernicus sentinel-1 and -2 data. *Remote Sens* 12:2824. https://doi.org/10.3390/rs12172824.

Robinson, I. S. 2004. *Measuring the Oceans from Space*. New York: Springer-Verlag.

Russ, J. C. 1992. *The Image Processing Handbook*. Boca Raton, FL: CRC Press, Inc.

Surowiecki, J. 2004. *The Wisdom of Crowds*. New York: Anchor Books.

Warner, T. A., M. Duane Nellis and G. M. Foody. 2009. *The SAGE Handbook of Remote Sensing*. London: SAGE Publications.

Weng, Q. 2014. *Global Urban Monitoring and Assessment through Earth Observation*. Boca Raton, FL: CRC Press.

Wilson, A. R., M. Vasilea, C. A. Maddock and K. J. Baker. 2022. Ecospheric life cycle impacts of annual global space activities. *Sci Total Environ* 834:155305. https://doi.org/10.1016/j.scitotenv.2022.155305

Woodhouse, I. H. 2006. *Introduction to Microwave Remote Sensing*. Boca Raton, FL: Taylor & Francis Group.

Index

Note: Page numbers followed by "f" denote figures; page numbers followed by "t" denote tables.